S0-BLS-267

STATISTICS IN SPECTROSCOPY

Second Edition

STATISTICS IN SPECTROSCOPY

Second Edition

H. MARK

Mark Electronics
69 Jamie Court
Suffern, NY 10901
USA

J. WORKMAN, Jr.

Argose Incorporated
230 Second Avenue
Waltham, MA 02451
USA

ELSEVIER
ACADEMIC
PRESS

Amsterdam Boston Heidelberg London New York Oxford Paris
San Diego San Francisco Singapore Sydney Tokyo

chem
01301894Z

Academic Press
An imprint of Elsevier
525 B Street, Suite 1900, San Diego, California 92101-4495, USA
http://www.academicpress.com

© 2003 Elsevier USA All rights reserved.

This work is protected under copyright by Elsevier, and the following terms and conditions apply to its use:

Photocopying
Single photocopies of single chapters may be made for personal use as allowed by national copyright laws. Permission of the Publisher and payment of a fee is required for all other photocopying, including multiple or systematic copying, copying for advertising or promotional purposes, resale, and all forms of document delivery. Special rates are available for educational institutions that wish to make photocopies for non-profit educational classroom use.

Permissions may be sought directly from Elsevier's Science & Technology Rights Department in Oxford, UK: phone: (+44) 1865 843830, fax: (+44) 1865 853333, e-mail: permissions@elsevier.com. You may also complete your request on-line via the Elsevier homepage (http://www.elsevier.com), by selecting 'Customer Support' and then 'Obtaining Permissions'.

In the USA, users may clear permissions and make payments through the Copyright Clearance Center, Inc., 222 Rosewood Drive, Danvers, MA 01923, USA; phone: (+1) (978) 7508400, fax: (+1) (978) 7504744, and in the UK through the Copyright Licensing Agency Rapid Clearance Service (CLARCS), 90 Tottenham Court Road, London W1P 0LP, UK; phone: (+44) 207 631 5555; fax: (+44) 207 631 5500. Other countries may have a local reprographic rights agency for payments.

Derivative Works
Tables of contents may be reproduced for internal circulation, but permission of Elsevier is required for external resale or distribution of such material.
Permission of the Publisher is required for all other derivative works, including compilations and translations.

Electronic Storage or Usage
Permission of the Publisher is required to store or use electronically any material contained in this work, including any chapter or part of a chapter.

Except as outlined above, no part of this work may be reproduced, stored in a retrieval system or transmitted in any form or by any means, electronic, mechanical, photocopying, recording or otherwise, without prior written permission of the Publisher.
Address permissions requests to: Elsevier's Science & Technology Rights Department, at the phone, fax and e-mail addresses noted above.

Notice
No responsibility is assumed by the Publisher for any injury and/or damage to persons or property as a matter of products liability, negligence or otherwise, or from any use or operation of any methods, products, instructions or ideas contained in the material herein. Because of rapid advances in the medical sciences, in particular, independent verification of diagnoses and drug dosages should be made.

First edition 2003

Library of Congress Cataloging in Publication Data
A catalog record from the Library of Congress has been applied for.

British Library Cataloguing in Publication Data
A catalogue record from the British Library has been applied for.

ISBN: 0-12-472531-7

∞ The paper used in this publication meets the requirements of ANSI/NISO Z39.48-1992 (Permanence of Paper).
Printed in Great Britain.

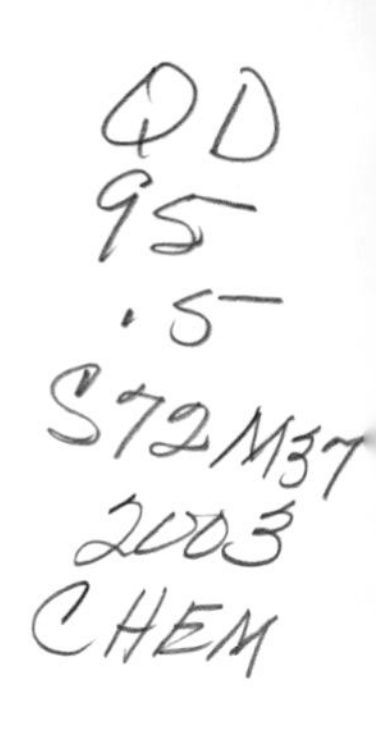
QD
95
.5
S72M37
2003
CHEM

To my wife Sara, my children Elyssa and Stuart, and
my grandchildren Joshua and Noah

—Howard Mark

To my wife Rebecca; and to
Cristina, Brian, Stephannie, Daniel, Sara, and
Michael (surf-on dude!)

—Jerry Workman, Jr.

CONTENTS

PREFACE

The second edition of this book is based on the series of columns "Statistics in Spectroscopy" that was running for several years in the journal *Spectroscopy*. The response to the column indicated that it was extremely popular and useful to the readers of the journal for describing sophisticated statistical techniques and terminology in an easy to read format.

When we initially considered writing the series it was in response to our having noticed how, for several years, the field of spectroscopy (at least, molecular spectroscopy) had departed from the "classical" approach. In this sense, spectroscopists have been relying less and less on their own personal interpretation of spectra in favor of using computers, usually with highly complex multivariate algorithms, to interpret the spectra for them.

The bases for these algorithms usually (although not always) arise out of the branch of mathematics called statistics. Whether or not the algorithm itself comes from that branch of mathematics, the data to which the algorithms are applied are invariably subject to random and non-random errors of various sorts; this is precisely the subject of the science of statistics.

Unfortunately, most chemists have not had the opportunity to learn about statistics or about how to apply statistical concepts properly to spectroscopy or to the design of experiments. Having discussed the matter among ourselves, we find that our own experiences are probably typical; due to limitations of time and curriculum we learned no more math than is required of a chemist, but after learning how statistics provides the tools to apply objective criteria to the relationships between real data and error (or noise) when used properly, our reaction was "How could any experimental scientist not know about this stuff?" Having spent more than twenty-five years between us before the publication of the first edition learning to apply statistical concepts to real spectroscopic situations, we had also heard others echo this same sentiment when they realized what statistics was really all about. At the publication of this second edition we have each spent an additional dozen or so years applying the concepts described in this book.

One reason statistics has been branded with a bad reputation is because it is sometimes difficult to use properly and has more pitfalls than any other area of science of which we are aware. Thus, results based on statistical calculations are sometimes (occasionally deliberately) misinterpreted and commonly used incorrectly or inappropriately. Surprisingly, even the most advanced chemists/spectroscopists (those developing the new algorithms) fall into the traps that await the unwary. We suspect that they get so enamored of the elegant mathematical transforms that they lose sight of the basics.

It was on this premise that we decided to present basic statistical concepts to the spectroscopic community using a tutorial format. After spending a good deal of time moaning and groaning over the situation as we saw it, and wondering when someone was going to do something about it (the "Let Joe do it" syndrome), we realized that we would have to be "Joe" or it wouldn't get done. Consequently, we proposed writing a series of articles to Heather Lafferty, the then editor of *Spectroscopy*. These articles were to cover topics relating the application of statistics to spectroscopic problems, starting from a very elementary treatment.

We therefore proposed to write such a column (or series) for *Spectroscopy* for the purpose of informing the spectroscopic community and dispersing this important information. We felt this would be an appropriate and effective way to get the word out. As a tutorial on statistics from the point of view of the spectroscopist, we could use examples based on data from actual spectroscopic results, as well as from synthetic data sets (when those better made our point).

Our goal in writing this book, as it was in the original series of columns, is to provide the general analyst or graduate student with a tutorial on the basic reasoning behind the proper use of the most common mathematical/statistical tools. We would also like to see it eventually used in undergraduate courses in analytical chemistry, so that students could start off the right way, but for the reasons mentioned in our introductory chapter, this may be too much to expect.

The mathematical/statistical tools discussed are those relating to spectroscopy and basic measurement science. The intent of the book is to avoid "heavy" mathematics, but to emphasize understanding and concept delineation. In this fashion, we hope to assist both the mathematically timid and the more advanced user. When results are better explained using derivations, or sometimes when the derivations are just plain "nifty" and therefore enlightening, they are included as an enhancement to the discussion. Another reason certain derivations are included is that the results are so startling that we feel that a rigorous mathematical proof is the only way to convince our readers that the results are true and correct, and also to convince our readers that statistics is indeed a mathematical science every bit as rigorous as any other subdiscipline of mathematics. When derivations are bulky and pedantic they are not used. We hope to provide the insights needed in order to use multivariate statistics properly, while avoiding heavy mathematics. Nevertheless, ensuring rigor is sometimes a necessary requisite for proper use of statistical results, therefore derivations may be required at some points. Most statistical derivations of interest can be done with no more than algebraic manipulations once the basic concepts are presented and understood. We therefore expect that no higher math will be needed for the vast majority of our discussions. In addition, there are a number of provocative "exercises for the reader" scattered throughout the text, as well as some computer programs (written in BASIC compatible with the IBM-PC version) that illustrate the points discussed.

There are several other reasons why we decided to rewrite the columns in book form. First, the columns are scattered throughout several years' worth of the journal; this makes it difficult for anyone who did not start reading the series at its inception and who did not save all the issues to obtain a complete set.

We often received a good number of requests for a specific column from the series- "and any other articles on this topic." It was not reasonable to try to provide the full set of columns under such conditions. We also received such requests from college teachers who wanted multiple copies of the series to distribute in their classes. While such action is desirable, indeed exactly to the point of our writing the columns in the first place, it is unfair to people we consider our benefactors, besides being illegal for us to do such a thing (to say nothing of the onerous burden that would place on us). Providing a full collection of the columns in book form addressed all these problems, as well as making the information available to some who might have missed the series entirely.

A third reason for writing this book is one that should be obvious to anyone who has ever published a book or article: the love/hate relationship that apparently exists between all authors and their editors. The various editors at Aster Publishing that we dealt with over the years were wonderful, yet we still had our share of fights with them over the editorial changes they made to our "babies." Sometimes we agreed with the changes, sometimes not; sometimes we won the fight, sometimes we lost-but how often does an author get a second chance to print his brainchild the way it was originally conceived, or at least get a chance to contend with a *different* set of changes, and try to wrest it back into the original form?

By writing the second edition of this book we also get a chance to add material clarifying and expanding upon certain topics, and it allows us to correct some minor errors that crept into the first edition. While this second edition stays very close to the original edition's content, format and style, it also gives us an opportunity to provide some new interpretations and points of view that were not in the original.

Howard Mark
Jerry Workman

1

Introduction: Why This Book?

The idea for this book came from a series of articles that themselves grew out of a group of discussions we had about the errors of interpretation of numeric data that appear in the chemical/spectroscopic literature from time to time, due to lack of knowledge of how such data should be correctly treated. Our major complaint was the usual "someone oughtta do something". Eventually, we came to the conclusion that we could either keep on complaining, or become the "someone".

For the past several years the field of spectroscopy (or at least molecular spectroscopy) has been departing from the "classical" approach. In this sense, spectroscopists have been relying less and less on their own personal interpretation of spectra in favor of using computers, usually with highly complex multivariate algorithms to interpret the spectra for them. Examples of these changes are the increasing presence of "Beer's Law", "inverse Beer's Law" and "factor analysis" in mid-infrared and FTIR spectroscopy, and "multiple regression" and "principal components" analysis in near-infrared spectroscopy. The availability of very powerful "micro"computers, together with the proliferation of sophisticated computer programs that implement the most complex of mathematical operations and bring them within reach of most practicing scientists has created a situation virtually unprecedented in analytical chemistry: scientists tend to use the available software without a full understanding of the meaning and, more importantly, the limitations of the results. Thus, sometimes there is a tendency to "try another algorithm" if the results were not quite what were expected, whether what was really needed was another algorithm, or better understanding and proper use of the existing algorithm.

One of the current trends in modern analytical spectroscopy is the incorporation of highly sophisticated mathematical algorithms into the computers and instrumentation used to extract useful information from raw spectral data. While methods such as principal component analysis, partial least squares and even such a "simple" technique as multiple regression analysis are founded in the science of statistics, few chemists or spectroscopists are aware of the basis of these algorithms, or proper methods of determining the meaning of the results produced. This statement is not intended to fault either chemists or our educational system. The controversy that appears periodically in *C&E News* over whether more curriculum time should be devoted to "basic principles" or to "keeping up with current technology" highlights the fact that there is simply not enough time to learn all the chemistry we would like to know, let alone other such "far-out" fields as Chemometrics or Statistics. Thus, there is a real danger of misuse of the tremendous computing power available, and it is all the more important to bring the topic of the proper use of statistical tools to the forefront.

Having discussed this subject with others, we find that our experience is typical; as "chemists" we learned no more math than required. However, having had the opportunity to learn what statistics is all about, and how to apply objective criteria to the relationship between real data and error (or noise) our reaction was, "How could any experimental scientist not know about this stuff?" Yet the rapid expansion of applied mathematics and chemometrics can also intimidate the practicing spectroscopist and hinder his use of these important mathematical tools.

The science of statistics has been branded with a bad reputation. This is due in part to the fact that it has more pitfalls than any other area of science that we are aware of, and is sometimes used incorrectly or inappropriately, thus leading to incorrect conclusions. Sometimes unscrupulous workers deliberately misuse statistical results, or use these results to confuse rather than enlighten (more of this long around Chapter 30).

However, the science of statistics is a valid field of scientific endeavor. It is rigorously based on mathematical principles, and, when applied correctly to problems for which it is the appropriate tool, can shed light on situations that are otherwise indecipherable.

Our goal in writing this book is to provide the chemist/spectroscopist with a tutorial on the basic mathematical/statistical tools needed to use statistics correctly and to properly evaluate situations to which statistical concepts apply. Our intent is to avoid heavy mathematics; thus, we hope to assist those who are mathematically "timid" to gain confidence in their ability to learn and to use these concepts. Some results are better explained when derivations are included, and therefore appropriate derivations of statistical ideas and methodologies will be included when, in our opinion, they will enhance the discussion. Mathematical derivations of statistical interest seem to fall into two categories. The first category involves the use of very advanced concepts from calculus and matrix algebra. We will avoid these. However, most statistical derivations that are of interest fall into the second category: they require no more than algebraic manipulation of a few basic concepts. There are few derivations of interest to us that fall in between and therefore we expect that no higher math will be needed for the vast majority of our discussions.

We also will try to relate statistical considerations to various situations a spectroscopist may encounter. Often, as happens in many other areas of intellectual endeavor, a lack of knowledge of a scientific discipline makes one unaware of the ways one's own problems could be solved using results from that discipline. Thus, while statistics, as a mathematical science, is a very generalized subject, we will use spectroscopic examples to show how statistical considerations can be applied to actual cases. Our own field of spectroscopic interest is near-infrared reflectance analysis, hence the majority of examples using real data will come from this sub-discipline. We will also often use computer-generated synthetic data sets to illustrate certain points. Such synthetic data can be made to have controllable and known properties, so as to simplify discussion by avoiding complicating factors that are often present in real data.

Pedagogically, using the series of columns as the basis for this book leads to some difficulties in regard to continuity. Sometimes, we will have to introduce some statistical ideas in isolation, in order to explore them in the desired depth, before tying them together with other topics, or with the spectroscopy involved. We ask our readers to bear with us and wait out discussions which may not seem immediately pertinent.

Why Statistics?

Before we can answer this question we must first consider the question of what statistics is. Statistics is very closely related to probability theory and, indeed, is derived from it. Statistics is, essentially, the study of the properties of random numbers. When we see what we call "noise" on a spectrum, we are seeing the random contribution of various uncontrollable phenomena to the net signal that we measure. Such random phenomena have characteristics that are definable and measurable, and knowledge of the properties of random numbers tells us what to expect about the behavior of the measurements that we make. Such knowledge also tells us how to design experiments so as to either minimize the effect of a given source of error, or at least to measure it, if the source cannot be controlled.

The science of statistics also serves auxiliary functions. Among these is the design of experiments that serve to utilize resources in an efficient manner while simultaneously minimizing the effect of noise (error) on the measurements.

Thus, a knowledge of statistics is important. Whether we use a complicated algorithm derived from statistical consideration, or perform a simple calculation on a measurement, or even just make a measurement, the data to which these algorithms are applied are invariably subject to random and non-random errors of various sorts; this is precisely the subject of the science of statistics. Many papers appear in the literature, including some by the developers of the new mathematical techniques, that draw unwarranted conclusions from inadequate data because the workers involved did not take proper cognizance of the fundamental statistical basis of their algorithms. Perhaps some of these workers get so enamored by the elegance of their mathematical structures that they forget to test or even consider the fundamental underlying assumptions upon which their work rests.

Aside from these more complicated situations, there are many simple situations that statistical considerations can be applied to, and many questions that arise during the ordinary course of spectroscopic work that can be answered by applying statistics.

For example, sooner or later every spectroscopist encounters a situation such as that shown in the digitized spectrum of Figure 1.1 (note that identification has specifically been removed, so that the spectrum can be considered "generic"): does the part of the spectrum marked with an arrow indicate the presence of a band or is it an artifact of the noise?

Figure 1.2 shows another situation, encountered when a calibration line is to be determined: in Figure 1.2-A there is clearly no relationship between the two

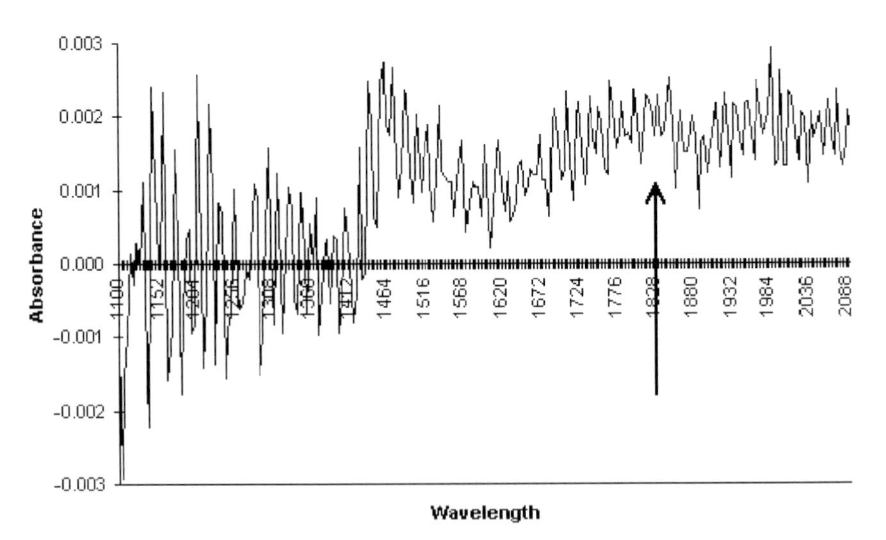

FIGURE 1.1. A section of a (generic) spectrum. Does the marked section of the spectrum indicate the presence of a band?

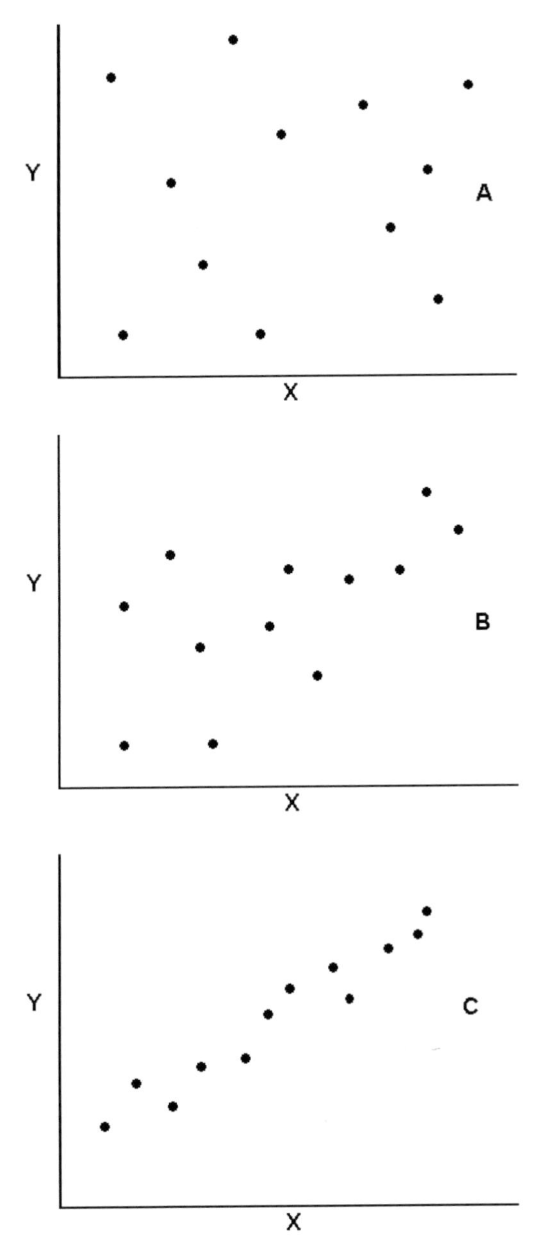

FIGURE 1.2. Different amounts of association between two variables. In cases A and C it is clear whether there is a relationship between the variables. Case B is marginal; it is not clear from the data whether a relationship exists.

variables being plotted while in Figure 1.2-C there is a clear relation. On the other hand, Figure 1.2-B represents a marginal case (and there is *always* a marginal case): is the relationship real or apparent, and how much confidence can we place in our decision, whichever way we decide? Even in case C, where one hardly needs to do a calculation to tell that a relationship exists, calculating numerical values for various characteristics of the graph can yield useful information. For example, if the abscissa represents instrument readings, and the ordinate represents reference values, then an indication of the precision and/or accuracy of the instrumental method may be obtainable.

Other questions arise that can be of either practical or intellectual interest:

How many times must I repeat an experiment in order to ensure reliable results?

Are two spectra "the same"?

What is the meaning of the published noise level of my instrument?

Why do we use the least squares criterion to calculate calibration lines?

How can I determine the contribution of any given error source to my final answer?

The list can go on and on. Each of these questions is answerable by proper application of the science of statistics to the data involved, which will then provide us with objective answers to these questions. However, we must note that these answers will not come out as a simple "yes" or "no". Rather, the answer will be a probabilistic answer. Furthermore, it is our responsibility to ask the questions in such a manner that we can decide on the correct answer by virtue of the probability of that answer becoming high enough.

The methods, and their proper sphere of application, were developed, in most cases, many years ago by statisticians of what we might think of as "the old school", i.e., before computers were sufficiently widespread that they could be used on a routine basis. These statisticians, of necessity, became very savvy in the ways of looking at data, and how to extract maximum information from any set of measurements. They were, if we allow the phrase, masters of "the Art of Science" and to some extent, have passed their knowledge on to us, but it is up to us to use it wisely. By comparison with these artists of data analysis, we will do well to become adequate "kitchen painters". Nevertheless, it is quite sufficient to be a good "kitchen painter", at least in the sense that learning the fundamentals of statistics, and when and how to apply them properly is something that anybody can do, and to their benefit.

2

IMPORTANT CONCEPTS FROM PROBABILITY THEORY

As we have discussed in Chapter 1, statistical concepts are derived from probability theory and usually result in probabilistic answers; therefore, it is useful at this point to review some of the fundamental meanings and properties of probabilities. Readers interested in pursuing this topic beyond our discussion here may wish to consult Reference (2-1), a book that covers the ground in considerably more detail than we can do here, but stays at the algebraic level of discussion that we intend to adhere to.

One may question the relevance of probability theory to spectroscopy. Indeed, the connection is rather indirect; our own initial reaction to this chapter after writing it was concern that our readers will wonder whether we are more interested in gambling than in spectroscopy. There is cause for this concern: consider that the initial impetus for developing probability theory was just to explain the phenomena seen (and calculate the odds) in gambling situations. It was not until later that the realization came that other natural phenomena could also be described, explained and calculated from probability theory.

For example, consider Figure 1.1 in Chapter 1. We will eventually learn how to calculate the probability of the individual data points having their measured values and, from that, the probability of the collection of points representing a spectral peak. From this we will conclude that there "probably is" a peak at the indicated location, or that there "probably is not" a peak. We may also conclude that we still cannot decide the question. However, in this last case we will at least know that the question is fundamentally undecidable.

So, let us start. The probability of an event may be defined as the fraction of the times that an event will occur, given many opportunities for it to occur. A common way that this concept is expounded is to present an imaginary jar filled with balls of different colors and perform the (imaginary) experiment of reaching into the jar and removing a ball. Clearly, if the experiment is performed many times, then the number of times a ball of some given color (let us say white) will be selected will be some fraction of the number of times the experiment is performed.

What the fraction will be may not always be easily determined before actually performing the experiment. However, in simple cases such as a jar full of colored balls, we can easily tell *a priori* what the probability of drawing any particular color will be. In many cases, particularly in real-world situations, such an *a priori* determination is not easily made.

For a jar full of colored balls it is intuitively obvious, and in this case intuition has been confirmed by both theory and experiment, that the fraction of times a (say) white ball will be selected will be the same (in the long run) as the fraction of balls in the jar that are white. We will also use the term "trial" to refer to performing an experiment or setting up any situation that allows for an outcome to either occur or not occur.

This definition of probability ignores many of the implicit properties of probability theory. Several difficulties involving the proper use of statistics are complicated by just such unstated factors. We note this fact here, even though we defer discussion of all the potential difficulties. Thus, while we must wait until a later chapter for a fuller discussion of some of the implicit factors, we note that the points of concern not stated in the definition are:

(A) requirement for randomness;
(B) nature of the distribution;
(C) probability versus expectation;
(D) discrete versus continuous distributions;
(E) meaning of "the long run";
(F) finite versus infinite distributions;
(G) finite versus infinite number of items.

We will consider only items A and D here.

The requirement for randomness seems obvious, yet is often overlooked. Too often, scientists working with numerical data will claim that errors of the data are normally distributed, whether they actually are or not (which usually is not known). Yet a normal (Gaussian) distribution is not critical to many statistical calculations, while randomness is.

What, then, constitutes randomness? Unfortunately, a simple yet mathematically correct answer does not exist. While we will return to the topic in Chapter 24 and discuss the requirements for a correct definition of "random", we will stick to the simple approach for now. As usual in mathematical logic systems, there must be some starting point that is acceptable, because no further pushing back of the levels of definition can be done. We are at that point here. Answers can be given by reference to other mathematical constructs, but these are not only beyond the level of mathematical depth to which we wish to keep the discussion, but also lie outside the scope of our current objectives.

Therefore, we will give an answer that is not mathematically correct, in order to help support our intuition as to the meanings of these terms. Indeed, our answer will be circular. With that proviso, we then can define randomness in terms of our "jar of balls" experiment: the selection of a ball from the jar is random if the selection is made in such a way that each ball has the same probability of being selected as any other ball. Suppose, for example, a jar contained 100 black and 100 white balls, and we draw a ball. Is the draw random? Perhaps, but suppose the jar also contained a partition, and all the black balls were below the partition. Now what is the probability of drawing a ball of either color? We like to think that such uncontrolled factors have been eliminated from our experiments, but it always pays to check.

Now, all this is fine for jars full of colored balls, but we do not usually deal with that. We usually deal with samples, measurements, computations on the measurements, and errors encountered in performing the measurements. How does our discussion relate to those things? Indeed the answer to that question will, in various forms, occupy our attention for many of the further chapters. A key difference between jars of balls and the results of instrumental measurements is that the result of a measurement is a selection of one reading from a potentially infinite set of possible values. Thus, for our immediate purposes, however, we can note that the relationship between jars of balls and errors in measurements is another form of point D above, insofar as the generation of the concepts of probability is made on a set of discrete items, while our use of these concepts are related to quantities that can take a continuum of values.

If we ignore the limited number of significant figures in a measurement and assume that we can make measurements of "infinite" precision, then it is clear that any measurement we make will have some particular value (we hope close to the "true" value for that measurement). In such a case, it is clear that what we actually obtain is one value selected from an infinite number of possible values that lie close to the "true" value. Thus, if we ask a question similar to the one in the finite case: "how often will I select (measure) a value of *nnn.nn*...?" then the answer is that I would *never* expect to get that particular exact value. Thus, the probability of making any one particular measurement is zero!

In the case of continuous data (as opposed to discrete) we need to do something else. What is done in this case is to redefine the nature of the "event" we wish to know about. Rather than the event being the occurrence of a single, given number, the "event" of interest is the occurrence of a number within a given predefined range of numbers. With this new definition, we can again count the number of times the event occurs, out of a known number of trials, and thus define probability for the case of continuous data also.

With the concept of probability thus suitably defined for the two cases, we find that certain properties exist. First, we will generate some nomenclature. We will use the symbol "P" as the label for the probability of an event occurring (we note that the "event" of interest in the probabilistic sense may be the *non*-occurrence of some physical event). Because P is the fraction of times an event occurs, then if a event occurs n times out of n trials, i.e., the event *always* occurs, then:

$$P = n/n = 1 \quad \text{(for an event that } \textit{always} \text{ occurs)} \tag{2-1}$$

Similarly, for an event that *never* occurs:

$$P = 0/n = 0 \quad \text{(for an event that } \textit{never} \text{ occurs)} \tag{2-2}$$

We now present three fundamental laws of probabilistic calculations for determining the probability of combinations of events (without proof; for the proofs, see Reference (2-1), pages 59–64):

(A) For mutually exclusive events (i.e., out of several possibilities only one of them can occur, but it is possible for none to occur)

$$P = P_1 + P_2 + P_3 + \cdots \tag{2-3}$$

where the P_i represent the probabilities of the individual events.

In words, the probability of one of several mutually exclusive events to occur is the sum of the individual probabilities.

Example 1: Draw one card from a standard deck (imaginary decks of cards are even better than jars of balls for these imaginary experiments!). What is the probability of it being either the four of clubs or the two of diamonds?

Ans: These are mutually exclusive outcomes, each has a probability of 1/52, therefore the net probability is 2/52.

(B) For all of a set of independent events (i.e., the outcome of one has no influence on another):

$$P = P_1 \times P_2 \times P_3 \times \cdots \tag{2-4}$$

In words, the probability of all of them occurring is the product of their individual probabilities.

Example 2: A card is drawn from a deck, replaced, then another one is drawn. What is the probability that both the four of clubs and two of diamonds were drawn?

Ans: The two draws are independent; therefore, the net probability is $(1/52) \times (1/52) = 1/2704$.

(C) For a set of *dependent* events, if the probability of the *i*th event is P_i only if the outcome of the $(i - 1)$th event (with probability P_{i-1}) is successful, then the probability of all the events occurring is:

$$P = P_1 \times P_2 \times P_3 \times \cdots \tag{2-5}$$

Example 3: A card is drawn, and a second card is drawn only if the first card is the four of clubs. Now what is the probability of getting those two cards?

Ans: The probability of drawing the four of clubs is 1/52, but if it is, then the probability of drawing the two of diamonds is 1/51. Thus, the net probability is $(1/52) \times (1/51) = 1/2652$.

Note that Equation 2-5 holds only for the dependent case as described. If, in Example 3, a second card is drawn regardless of the outcome of the first draw, the situation becomes considerably more complicated. This complication arises from the fact that it becomes necessary to take into consideration the different probabilities that arise if the two of diamonds card had been drawn on the first draw, if the four of clubs had been drawn, or if neither one had been drawn.

From these theorems, we can obtain other results that are not only interesting, but also that we shall find useful in the future. These derivations will also illustrate methods for performing probabilistic calculations.

(D) We will now define the symbol "$\sim P$" as the probability of the non-occurrence of an event. In our universe of probabilistic discourse, if a trial is made, then an event must either happen or not happen. Because these two outcomes are mutually exclusive and it is certain that one of them occurs, we can apply Equation 2-3 to obtain the result:

$$P + \sim P = 1 \tag{2-6}$$

(E) For a set of independent events each with probability P_i to occur, what is the probability of *any* of them occurring? (Note that Equation 2-4 is the probability of *all* of them occurring.)

Step 1: In the form given, the question is not answerable directly. However, it is possible to calculate $\sim P$; then from Equation 2-6, $P = 1 - \sim P$ and we can ultimately find *P*.

Step 2: The symbolism $\sim P$ means that none of the events may occur, or, in other words, *all* the events must *not* occur; then we can apply equation 2-5 thusly:

$$\sim P = \sim P_1 \times \sim P_2 \times \sim P_3 \times \cdots$$

Step 3: Each of the $\sim P_i = 1 - P_i$ (again from Equation 2-6), therefore:

$$\sim P = (1 - P_1) \times (1 - P_2) \times (1 - P_3) \times \cdots$$

Step 4: Substitute this expression into the expression in Step 1:

$$P = 1 - ((1 - P_1) \times (1 - P_2) \times (1 - P_3) \times \cdots) \tag{2-7}$$

Example 4: A card is drawn, replaced, then another one drawn. What is the probability of having drawn the four of clubs at least once?

Ans: The two draws are independent, in each case the probability of drawing the desired card is 1/52. Therefore, the total probability is:

$$1 - (51/52) \times (51/52) = 1 - (2601/2704) = 103/2704 \text{ (approx. 0.0381).}$$

Example 4 points up an important special case of equation 2-7: If the individual probabilities P_i of any of several events are all the same, then the probability of any of them occurring is:

$$P = 1 - (1 - P_i)^n \tag{2-8}$$

where n is the number of individual events.

There is another form of dependency that can occur in probabilistic calculations; that dependency is such that the probability of an event changes with the outcome of previous events. We have seen this in Example 3 above. As an "exercise for the student", consider the following problem:

Draw the top two cards from a shuffled deck, in order. What is the probability that the *second* card is the four of clubs?

Hint: If the top card is the four of clubs, the probability of the second card being it is zero. If the top card is not the desired card, the probability of the second card being correct is 1/51.

Ans: The total probability is 1/52. We provide the answer because the real problem is to prove it, using the previous considerations.

Note that in all our examples, similar situations were set up, involving hypothetical draws of cards from an imaginary deck. The differences in the situations lay in the details: how many cards were drawn, were they drawn in order, was a drawn card replaced, were the draws independent, how did prior knowledge affect our computation of probabilities, etc. An important point needs to be made here: The differences in the details of the situation changed the computation required to obtain the correct answer. Obviously, it is not possible to provide formulas to perform the calculations for all possible cases, because the number of possibilities grows combinatorially as different factors are considered. These considerations, of course, extend to more complicated situations, particularly when we come to apply statistics to real-world situations. It becomes necessary to be intimately familiar with the fundamental properties of the situation, so that the correct calculation to use can be selected by combining the correct pieces (building blocks) of statistical knowledge. That way, the apparent complexities of needing to know dozens of formulas to cover all the different cases disappear.

For practice, we set up the following situation and ask a few questions that may or may not be answerable by applying the formulas listed above, i.e., you may need to go back to the definition of probability in order to compute the answer:

The situation is as follows: One jar (call this jar #1) of balls contains half-and-half (i.e., 50% each) black and white balls. Jar #2, which is twice as big as jar #1, contains 1/3 white and 2/3 black balls. The balls in the two jars are combined and mixed, and a ball is drawn.

Questions:

(1) What is the probability that the ball is white?
(2) What is the probability that the ball came from jar #1?
(3) If the ball was white, what is the probability that it came from jar #1?
(4) The ball is replaced and another one drawn. What is the effect (for the second ball) on the answers to questions 1–3?
(5) If a second ball is drawn without replacing the first, what is the effect on the answers? [Note that questions #4 and #5 cannot be answered numerically, and are presented to provoke (and promote) thinking about the subject (i.e., what needed information is missing?).]

The Connection Between Probability and Statistics

It is always harder to notice something missing than something present. That is why we are now going to point out something missing from the above discussion.

The missing item is a discussion of distributions. Except for one brief mention of a misuse of the normal distribution, we have avoided bringing the question of the distribution of data into our discussion. The reason for this is that the distribution of a set of data is important only insofar as it affects the probability of finding any particular value in the set of data at hand (which indeed is *very* important when determining the properties of the data); but the probability itself is the truly fundamental property.

Thus, when statisticians chose the various methods of calculating numbers to describe the properties of data (such as mean and standard deviation), these methods were not chosen arbitrarily, nor for convenience, nor even for tractability. One of the questions we presented in Chapter 1 of this book was, "Why do we use the least squares criterion to calculate calibration lines?" The answer to this question is the same as the reason for performing the other calculations that are used: it represents what is known as a "maximum likelihood estimator", i.e., the result is the number that is the "most likely" outcome, in a well-defined sense that we will not go into here. Such calculations also have other useful properties, which we will investigate further on.

This brings us to the point of discussing the nature of the fundamental statistical operation: what is known in statistics as a "hypothesis test". In various guises, the hypothesis test is the tool used to make statistical statements. The basis of a hypothesis test is to make what is known as a "null hypothesis". The nature of the null hypothesis will depend upon the data available and the information desired; a good number of chapters will eventually deal with the proper null hypothesis to make under different conditions. Sometimes an experimental situation will have to be modified to accommodate collecting the data needed to test the null hypothesis. A mistake often made is to collect data arbitrarily, or at least without regard to the statistical necessities, and then wonder why no reasonable statistical judgments can be made. These necessities, it should be clear by now, are not arbitrary, but are the result of trying to obtain information in the presence of random phenomena affecting the data. An improperly designed experiment will allow the random factors to dominate; thus any measurement will be only of the noise. A properly designed experiment will not only suppress the random factor, it will allow an estimate to made of its magnitude, so that the contribution it makes to the measurement can be allowed for. If proper care is not used when designing the methods, procedures, and criteria for data collection, with due consideration for the effect of random variations superimposed on the data, reliable conclusions cannot be drawn. Indeed, the purist will, on the basis of the information desired, insist on setting up an experiment and forming the null hypothesis before collecting any data at all. Such rigor is not usually needed in practical cases. However, when applying statistics to data that has already been collected, it is important to act *as though* this had been done. Historical data (i.e., already collected) can give valuable insights into the nature of phenomena affecting the results of an experiment, but one must be careful because specifying a null hypothesis based on the examination of data already collected will distort the probabilities as the data can no longer be considered random. Forming a null hypothesis on the basis of a perceived characteristic of the data at hand induces non-random character to the data and probability levels calculated on the expectation of random behavior do not apply. "Seek and you shall find" applies here; if you test for a characteristic because you already think you see that characteristic in the data, you are more likely to find it with a hypothesis test than if the test were applied blindly: the foreknowledge affects the result.

Similarly, only certain operations are permitted on different statistics, in order to retain the correct probability levels. Performing inappropriate calculations on statistical numbers will give the same result as performing an inappropriate experiment: a number will result, but the meaning of that number will be impossible to determine.

Having formed a null hypothesis, the next step is to determine the probability that the data collected from the experiment at hand is consistent with the null hypothesis. If the probability is high, then the null hypothesis is accepted; if the probability is low, then the null hypothesis is rejected. However, the two cases are not symmetric: rejection of a null hypothesis is a much stronger conclusion than acceptance. The reasons for this will become clearer as we get deeper into the subject, but for now we point out that accepting a null hypothesis does not preclude the possibility of the existence of other null hypotheses that might also be consistent with the data. The formalism of statistics requires us, if the probability of data being consistent with the null hypothesis is high, to phrase our conclusion as: "*There is no evidence* to justify rejecting the null hypothesis", to emphasize the weakness of the conclusion. Thus, the usual procedure is to design the experiment (if possible) so that the desired information can be gained by *rejecting* the null hypothesis.

The marriage of statistics and probability occurs when, after observing a phenomenon, we compare our *actual* results with the *expected* results. Probability theory allows us to calculate the results expected from some model of the system of interest, while statistics enables us to determine whether our data is consistent with that model, and thus forms a basis for determining whether the model is valid. Statistics thus gives us a critical and sensitive numerical measure of disagreement between observed and expected results. The term "statistically significant" is used to indicate that the data at hand is so inconsistent with a postulated model that it would have been virtually "impossible" for that data to have been obtained if the model was correct, allowing for sources of error, deviation, variation, etc.

The determination of whether or not the real data is in disagreement with the expected results is made by calculating some "test statistic", which could be a mean, standard deviation, Student's t, Chi-square, normal distribution, F-distribution, or any of a large set of others. The value of the test statistic is compared with the values for that statistic known to correspond to different probability levels. At some probability level, known as the "critical level", we decide that our data is too unlikely to have occurred if the model was correct.

We must keep in mind that we may occasionally reject a true null hypothesis, or vice versa, accept a false one. The probabilities of making these mistakes can be changed by adjusting the critical level. Ideally, we would like to find a critical value that would reduce the probability of both types of mistake to an acceptably small value. Such a situation can occur only if the data is suitable. Otherwise, we can reduce the chance of making one mistake only by increasing the other. On the other hand, just knowing this is enough to let us decide if the data is suitable, or to design our experiment to make sure that we collect suitable data.

Just as chemistry is divided into sub-disciplines, so is the science of statistics. Of interest now is the fact that one way to divide statistics is analogous to the separation of chemistry into "pure" and "applied": the corresponding branches of statistics are called "mathematical statistics" and "applied statistics". We have noted before that we plan to avoid deep mathematical expositions. This is possible because we will discuss mainly the branch called "applied statistics". Mathematical statistics is the branch dealing with the derivations of the expressions to use; applied statistics is, as the name implies, that branch that deals with the proper use of statistics to solve problems. Thus, by avoiding the details of many of the derivations we can avoid heavy mathematics. We will, however, occasionally find it useful to discuss the derivations qualitatively to gain some insight as to what is going on.

Statistics and probability theory allow explanations for the nature of errors in quantitative analysis, determination of the expected variations in spectra when repeating scans of different aliquots of the same sample, resolution of real or perceived differences in performance between quantitative calibration or qualitative spectral matching algorithms, and quantitation of differences in samples, spectra, and instruments.

We will return to these topics after discussing some other fundamental statistical concepts: the concept of a population, a statistical sample (as opposed to a chemical sample) and the notion of degrees of freedom.

References

2-1. Alder, H.L., and Roessler, E.B., *Introduction to Probability and Statistics*, 5th ed., W.H. Freeman and Co., San Francisco, 1972.

3

Populations and Samples: The Meaning of "Statistics"

The statistical notion of a sample is closely related to, although distinct from, the chemical meaning of the word sample. To the chemist the word "sample" conjures up the image of a small pile of something, or perhaps a small container of liquid, paste, or maybe "gunk", for which he is to determine the composition in some manner. The statistical meaning of the word "sample" is more closely allied to the chemist's term "representative sample".

A "representative sample" implies that there exists, somewhere, a large pile (or container of gunk) for which someone wants to know the composition but, for any of a number of reasons, the large pile cannot be analyzed. Presumably an analysis of the small pile will tell what the composition of the large pile is.

Populations

To the statistician, the large pile corresponds to the statistical notion of a "population". A population is a collection of objects that all share one or more characteristics. The key point about a population is that the collection is complete, i.e., a population includes *all* objects with the necessary characteristics. It should be emphasized here that the "objects" need not be *physical* objects. A set of measurements, or the errors associated with those measurements, can form a valid statistical population, as long as references to those measurements/errors are intended to include all cases where that measurement is made.

While the characteristics defining a population need not be quantitative, from the statistical point of view we generally are interested in those characteristics of the objects in the population that can be described numerically. For example, chemical measurements can be divided into two categories: qualitative and quantitative. Thus, a chemical specimen (terminology recommended by IUPAC, to avoid any possibility of confusion over the use of the word "sample") can have non-numerical characteristics, e.g., color, crystalline form, etc., as well as characteristics that can be determined quantitatively (percent composition, density, etc.) Similarly, the human race can be divided into two populations: men and women, each population having suitable defining characteristics. Items belonging to both populations can be seen to have qualitative (e.g., hair color) and quantitative (e.g., height and weight) characteristics.

Parameters

In some cases it is possible to describe those characteristics of interest in a population by tabulating the values for all the members of the population. In principle this could be done even if the size of the population is very large, as long as the population is finite. However, some populations are not finite, particularly populations that are described by mathematical definitions. Thus, in order to describe the characteristics of an infinite population, and for convenience in describing finite but large populations, it is necessary to summarize the characteristics of interest. These summaries are generated by performing calculations on the values of the individuals comprising the population, in order to condense the information into a few numbers.

These numbers, which describe the characteristics of interest in a population, in proper statistical nomenclature are called the "parameters" of the population. Parameters commonly calculated for many populations include the mean (arithmetic average) and standard deviation (a measure of spread). Since parameters represent the value of the corresponding characteristic for the whole population, they are known exactly for that population, and cannot change. However, some populations change; for example, births and deaths continually change the populations of both men and women. If the population changes, how can we define a parameter? The answer to this apparent dilemma is to make a suitable definition of the population of interest, and failure to define the intended population properly is another one of the pitfalls in the path of proper use of statistics.

In the case at hand, for example, we can define the population of men, or women in either of two ways. One definition would include in the population all those and only those who are alive *right now*. Another possible definition would be to also include all men (women) who were ever alive, or will ever be alive. Whichever way we choose, we must then stick to that definition: no fair changing the definition of the population in the middle of a calculation! In either case, once the population is defined, the values of the parameters for that population are completely fixed.

Samples

A sample, in contrast, is a suitably selected subset of the population. Usually, the selection is made in a random manner, so that, as we have noted before, all individuals in the population have an equal chance of appearing in the sample. Sampling is often done for convenience, because the number of individuals in the population is simply too large to evaluate all of them. In some cases, such as the second definition of populations of men/women above, it is not possible to evaluate all the individuals of the population, because some of those individuals do not exist yet. In such cases, we wish to select a sample from the population that has the same properties as the population from which the sample is drawn. Then, if we perform the same calculations on the sample as we would have performed on the population, we should get the same result, and thus characterize the properties of the population by calculating the properties of the sample.

A spectrum, for example, is also a sample. Indeed, it is very complicated because it is several samples rolled into one. One way to view a spectrum of some substance is as a sample of size one selected from the population of all possible spectra of that substance. If the property of interest is the characteristics of the instrument used to measure the spectrum, then the spectrum can be considered a sample of size n (where n is the number of wavelengths measured, for an optical spectrum) of the noise of the instrument making the measurement. A third view of a spectrum is as a sample of size m (where m is the number of impurities with measurable absorbances) selected from the set of all possible spectra of the substance with m impurities, where the m impurities are a subset of all possible impurities.

It is even statistically valid to consider a spectrum as a population, where the defining characteristic of the population is that it represents the set of numbers collected from a particular sample in a particular way at a particular time. While valid, this approach is not very useful, because it precludes generalizing any results obtained to other spectra of any sort.

At this point we run into two problems. One is near-trivial, but we will discuss it in order to avoid confusion in the future. The other problem lies at the heart of the science of Statistics, and, indeed, is its reason for existence.

Statistics

The trivial problem is one of nomenclature. We have been using the word "Statistics" to describe the name of the mathematical science that concerns itself with the properties of random numbers, and to the analysis or interpretation of data with a view toward objective evaluation of the reliability of conclusions based on the data. There is also another meaning for the word "statistics"; this second meaning is used in discussions of results covered by the science of Statistics. In this second sense, the word "statistics" is used to describe the results of calculations on samples that correspond to the calculations that result in a parameter when performed on a population. For example, the mean height of all women (a population) is a parameter. However, the mean height of some group of women less than *all* women (a sample) is a statistic. It is unfortunate that the same word has two different meanings within the confines of a single science, and may upset the purist, but that is the way it is. When used, the meaning is usually clear from the context, but we will also distinguish between the two by using the word "Statistics" (capitalized) to refer to the science of Statistics, and the word "statistics" (lower case) to refer to calculated properties of a sample.

In one sense, statistics are to samples what parameters are to populations. However, there is an important difference, one which brings us to the critical problem, the one that lies at the heart of the science of Statistics, that we approach with a computer experiment. We define a population of numbers (that being about the only population you can get on a computer), and then investigate its properties. Appendix 3-A is a listing of the program we used. We provide it so that anyone can run this program on his own and perform his own investigations on this topic. This program is written in Microsoft BASIC and was run on an IBM PC-AT, but should work on virtually any computer with this version of BASIC, and most versions of BASIC.

On line 450 of the program we define the population: it is the sequence of integers from 1 to 10. We use this as our test population because it is easily defined, it is small enough to calculate the parameters easily and large enough to provide intuitive confidence that the calculations are meaningful. The mean of this population is easy to compute and has the value 5.5.

The program then selects 20 sets of random samples from the population and computes the mean of the samples. The size of the sets can be varied (by rewriting line 10). At this point, in order to progress we will simply assume that the computer function RND produces numbers that are sufficiently random for our purposes, although we will eventually discuss ways of testing this assumption. Statement 10 of the program specifies how many items will be taken from the population and averaged. Table 3-1 presents the results of running this program, showing the results from not averaging ($n = 1$) and from averaging between 2 and 100 samples. In each case the experiment is repeated 20 times.

This simple computer experiment illustrates some important properties of samples. First, hardly any of the averages of the samples, shown in Table 3-1, are the same as the population average. While Table 3-1 contains results from 140 sets of samples drawn from the population, only three of these have averages that are the same as the population average of 5.5.

TABLE 3-1
Each column contains 20 means of random samples from a population of 10 integers (1–10), where n samples were used to calculate each mean. Note: the population mean of the integers from 1 to 10 is 5.5.

$n = 1$	$n = 2$	$n = 5$	$n = 10$	$n = 20$	$n = 50$	$n = 100$
4	6	6.8	5.5	5.3	5.24	4.96
8	6	4.2	5.1	5.45	4.68	5.32
7	8.5	5.8	5.4	4.8	5.1	5.38
5	3	4.4	5.5	5.3	5.54	6.06
10	4	4.4	4.7	3.95	5.22	5.7
7	6.5	6.4	4.9	5.3	5.54	5.69
2	3.5	5.4	7.4	4.8	5.76	5.72
4	7.5	5.6	3.2	5.2	6.36	5.68
6	3	4.4	3.7	5.4	5.06	6.07
2	5	5	4.2	5.9	6.34	5.22
10	5	4.4	3.8	5.3	5.56	5.39
3	4	5.4	6.8	5.3	5.82	5.58
4	6.5	6.8	5	5.2	5.92	5.48
3	7	8	4.6	6.45	5.52	4.81
9	4.5	3	5.3	4.65	6.3	5.68
6	6	3.4	5.1	5.8	5.06	5.11
5	4.5	3	6.2	5.9	5.92	5.53
1	7.5	4.4	4.6	6	6.22	5.46
9	5.5	4	5.4	6.75	4.8	6.06
1	4	4.4	6.4	5.85	5.64	5.86

A second important property illustrated by these results is that the sample statistics (the mean, in this case), tend to cluster around a certain value.

A third important property of samples shown in Table 3-1 is that, as the samples become larger and larger, the clustering becomes tighter and tighter. Inspection of the values in the various columns of Table 3-1 will show a decrease in the range as more and more objects are included in the average. Other measures of the dispersion of the results also show this tightening up of the values as more and more samples are averaged.

A fourth property shown in Table 3-1 is that the value around which the averages cluster is the population mean. This property of the mean is important, in that it is a property shared by many, but not all, statistics.

The property of a statistic tending to cluster around the population value for that characteristic is so important that there are several terms used to describe this situation and this leads us to a good opportunity to present some more statistical terminology. A statistic that tends to cluster around the corresponding parameter is said to "estimate" the parameter, and a statistic that estimates the corresponding parameter is called an "unbiased" statistic (sometimes "unbiased estimator").

The purpose of collecting samples is to estimate the population parameters since that is the "long run" value we want to determine.

For the converse case, where a statistic clusters around some value other than the parameter, the statistic is called "biased". Bias in another word that has more than one meaning within Statistics. For now we will simply note this fact without discussing the other meanings at this point.

In general, the goal is to use unbiased statistics, precisely for the reason that they estimate the population parameters. This is what allows statements to be made about a population's characteristics from the measurements on a sample. Since a biased statistic does not necessarily estimate anything in particular, it is simply the result of some computation on a set of numbers that may or may not mean anything, unless some bounds can be put on the bias. Unfortunately, in some cases unbiased statistics cannot be generated. While these cases are few, some of them are important, and we will discuss them at the appropriate time. For now we note that we always try to work with unbiased statistics whenever possible.

While on the subject of terminology, we will note one more point. It is necessary to distinguish between parameters and statistics when performing calculations. To this end, statisticians have adopted a convention: parameters are designated by Greek letters and statistics by Roman letters. For example, the mean is designated by the Greek letter μ when the population average is calculated, and by $\bar{X}$ when the sample statistic (average) is calculated. Similarly, standard deviation is conventionally designated with a σ for a population, and with the letter S for a sample.

Our simple computer experiment is an analogue of real-world physical experiments where data is subject to the random fluctuations ascribed variously to "noise" or "experimental error". While it is very simplified, it does share many important properties with the random errors that real data is subject to.

We will eventually discuss these facts quantitatively; for now it will suffice us to know that:

(1) A reading that is subject to random variability is never exact; only by chance do we attain any given result.

(2) Making a measurement on a sample results in a statistic that tends to be in a cluster around some value; if the statistic is unbiased (as the mean is) then it clusters around the population parameter for the same quantity. Of course, if only one sample is measured, the cluster is only conceptual. However, a statistic rarely, if ever, has exactly the same value as the parameter. Even more importantly, it will rarely, if ever, have the same value as another calculation of that same statistic from a similar, but different, measurement. This fact is often overlooked, but it a key point in properly interpreting the results of Statistical calculations.

(3) Increasing the size of the sample tightens up the cluster, so that the average from any large sample is likely to deviate less from the population value than one

obtained from a smaller sample. Note the use of the word "likely". Here is where probability theory starts sticking its nose into the results of calculations. We will eventually examine exactly *how* unlikely different deviations are.

Before we do that, though, we will examine, in the next chapter, the statistical notion of "degrees of freedom".

Appendix 3-A

Computer Program to Generate Means of Random Samples

```
10 N = 1 : REM number of items to average, change for
   different runs
15 DIM POP(10)
20 REM randomize timer :rem remove first remark to change
   numbers
30 FOR I = 1 TO 10: READ POP(I): NEXT I: REM get data (from
   line 450)
40 FOR I = 1 TO 20 : REM generate 20 averages
50 GOSUB 9000 :REM get an average of n numbers
300 PRINT XBAR :REM print it
350 NEXT I : REM end of loop
400 END : REM end of program
450 DATA 1,2,3,4,5,6,7,8,9,10
9000 REM
9001 REM subroutine to average n numbers from POP
9002 REM
9050 SUMX = 0 : REM initialize
9090 FOR J = 1 TO N :REM we're going to average n numbers
9100 GOSUB 10000 : REM get a random number
9120 SUMX = SUMX + A : REM generate the sum
9130 NEXT J :REM end of summation
9140 XBAR = SUMX/N : REM calculate mean
9150 RETURN
10000 REM
10001 REM subroutine to get a random population member
10002 REM
10010 A = RND(A) :REM generate random number
10020 A = INT(10*A) + 1: REM convert to an index
10030 A = POP(A):REM get a member of the population
10040 RETURN
```

4

DEGREES OF FREEDOM

What are degrees of freedom and why do we use them in order to perform statistical computations? To answer the first part of the question is actually fairly simple, although the answer will not seem to make much sense for a while.

The number of degrees of freedom in a set of data is the number of independent measurements or numbers constituting that set. For example, suppose we wish to choose a set of five arbitrary numbers. If we wish, we can choose the following set of five numbers:

3 5 17 2 10

Note that when we chose these numbers, we were able to choose any number we pleased at any stage of the process: We had five *degrees of freedom* for choosing numbers, and thus the numbers have, in them, those five degrees of freedom.

Now we change the scenario. We still wish to choose five numbers, but now we subject our choice to a constraint: the mean value of the numbers is to be 8.0. Now we can choose the first four numbers as we did before:

3 5 17 2 ?

The choice of the fifth number is no longer unrestricted: we are no longer *free* to choose any number for the fifth selection. Our freedom of choice for the fifth number has now collapsed, to the point where one and only one number will satisfy the conditions for choosing: the fifth number must be 13. *We have lost one degree of freedom* from the numbers when we specified the mean.

Similarly, if we specify that the numbers we choose must have a specified mean and standard deviation, we restrict the choices still further. If we select the first same three numbers:

3 5 17 ? ?

and insist that not only the mean equal 8.0, but also that the standard deviation be 6.0, then we find that our choice of both remaining numbers is restricted; only by choosing:

3 5 17 3.725 11.275

can we satisfy the dual conditions. We have lost two degrees of freedom to choose, and the five numbers contain only three degrees of freedom because of the constraints. Similarly, were we to specify other conditions the numbers had to meet (e.g., they must have certain values of skew or kurtosis), we would find that our freedom of choice of numbers was further and further restricted.

So what does this have to do with real data? If we measure a spectrum the data representing the spectrum is not subject to predetermined, arbitrary restrictions on the values they might take. Indeed, such raw data does indeed contain as many degrees of freedom as we have independent measurements (one degree of freedom for each wavelength), just as in our first number choice where there were no restrictions.

However, when we start to manipulate the data, the number of degrees of freedom can change. A common calculation is to compute the mean of a set of numbers and subtract the mean from each of the data. Now, you see, we have indeed forced the numbers to have a predetermined value of the mean, the value happens to be zero. However, the value forced upon the mean is not important to our present discussion, what is important is the fact that there now exists a forced relationship between the data points. This relationship is such that the possible values for the data are no longer unrestricted. Out of n measured data points, only $n - 1$ are unrestricted. Knowing the values of those $n - 1$ points, we see that (and we could compute) the nth point is forced to have one particular value and no other because the restriction that the mean of all the data is zero. Thus, while the loss cannot be associated with any single one of the numbers, data that have had the mean subtracted have lost a degree of freedom, just as specifying the mean removed a degree of freedom from our arbitrarily specified set of numbers.

Similarly, calculating the standard deviation of a set of numbers and dividing each datum in the set by that value removes another degree of freedom, because knowing $n - 2$ of the results forces the values of the remaining 2; thus two degrees of freedom have disappeared.

This procedure generalizes to all statistical calculations: having calculated m statistics from a set of n data points, operating on the initial dataset with those statistics removes a degree of freedom from the data for each statistic used to operate on the data. The resulting data contains $n - m$ degrees of freedom. When performing statistical calculations, degrees of freedom are used in the calculations, because using this procedure to calculate further statistics causes those statistics to be unbiased estimators of the corresponding parameters. This fact is not at all obvious, so we shall spend the rest of the chapter discussing this point.

We shall use the calculation of standard deviation as our example calculation, for a number of reasons. This is a common calculation, everyone knows what it means [or at least thinks he knows this parameter as a measure of the dispersion (or spread) of a set of numbers]. It is the simplest calculation for which these considerations apply, and the results are easily (and we hope, convincingly) shown. Furthermore, for this case, rigorous derivations are readily available (see p. 231 in Reference (4-1)). Since rigor is available elsewhere, we will concentrate our attention on trying to understand what is going on, rather than present another, perhaps incomprehensible, derivation. This is a procedure that we will use often: we will discuss the simplest possible case of a procedure, in order to gain understanding, then we will accept the fact that proofs exist for the more complicated cases we will deal with.

The expression for calculating standard deviation is sometimes seen to have either of two forms:

$$\text{S.D.} = \sqrt{\frac{\sum_{i=1}^{n}(X - \bar{X})^2}{n}}, \qquad \text{S.D.} = \sqrt{\frac{\sum_{i=1}^{n}(X - \bar{X})^2}{n - 1}} \tag{4-1}$$

the only difference being the use of n or $n - 1$ in the denominator. Indeed, some pocket calculators provide buttons to give one result or the other, as desired. When do you use which form, and why?

Again, the answer lies in whether you are dealing with populations or samples. In a population, the standard deviation is a unique, fixed quantity, just as all other parameters are. Furthermore, a population has the full number (n) of degrees of freedom. Thus, the form of the expression for standard deviation, using n in the denominator, is defined as the standard deviation for a population.

We will again use our population of the integers 1–10, as introduced in the previous chapter, for our test population. The value of standard deviation for this population (designated σ) is 2.8722813... We modified the previous computer program to calculate the standard deviation of random samples from this population, just as we previously calculated means. The listing for this program is in Appendix 4-A, and the results of running this program, for various different numbers of random selections for the population are shown in Table 4-1.

There are many things that can be seen in Table 4-1, but for now we will concentrate our attention on the first two columns, the ones corresponding to $n = 3$, and using n and $n - 1$ in the denominator of the standard deviation

TABLE 4-1

Each column contains standard deviations of n random samples selected from the population of integers 1–10. Differences are computed around the means of the individual sets of samples. For each value of n, the standard deviation was computed using n or $(n - 1)$ in the denominator of the standard deviation equation. The population value of standard deviation for the set of 10 integers is 2.8723...

$n = 3$		$n = 10$		$n = 30$		$n = 100$		$n = 300$	
$(n-1)$	(n)	$(n-1)$	(n)	$(n-1)$	(n)	$(n-1)$	(n)	$(n-1)$	(n)
2.081666	1.699673	2.592725	2.459675	2.808035	2.760837	2.974368	2.959459	2.785221	2.780575
2.516612	2.054805	3.314949	3.144837	2.965123	2.915286	2.5101	2.497518	2.828969	2.82425
2	1.632993	2.75681	2.615339	3.255588	3.200868	2.85961	2.845276	2.775693	2.771063
4.358899	3.559026	2.953341	2.801785	3.139661	3.08689	2.80627	2.792204	2.815269	2.810573
3.21455	2.624669	3.164034	3.001666	2.266447	2.228353	2.805118	2.791058	2.855279	2.850516
2.645751	2.160247	3.034981	2.879236	2.4516	2.410394	2.887434	2.87296	2.857639	2.852872
4.041451	3.299832	2.1187	2.009975	2.649008	2.604483	2.803605	2.789552	2.858257	2.85349
2.309401	1.885618	2.973961	2.821347	2.738613	2.692582	2.863494	2.84914	2.884973	2.88016
2.886751	2.357023	3.020302	2.86531	2.854518	2.80654	2.667633	2.654261	2.962063	2.957121
3.511885	2.867442	2.859681	2.712932	2.865771	2.817603	2.706446	2.69288	2.798492	2.793824
1.527525	1.247219	3.29309	3.124099	2.887946	2.839405	2.813729	2.799625	2.914024	2.909163
4.041451	3.299832	2.616189	2.481935	3.22419	3.169998	2.937669	2.922944	2.964244	2.959299
3.511885	2.867442	2.708013	2.569047	2.397316	2.357022	2.931818	2.917122	2.927735	2.922851
1.154701	.942809	2.319004	2.2	3.05975	3.008322	2.794999	2.78099	2.891336	2.886513
4.163332	3.399346	1.888562	1.791647	3.085003	3.03315	2.792052	2.778057	2.868881	2.864096
1.527525	1.247219	2.469818	2.343075	2.284631	2.246231	2.94356	2.928805	2.940701	2.935795
4.358899	3.559026	2.936362	2.785678	2.490442	2.448583	2.720869	2.70723	2.905075	2.900229
3	2.44949	1.776388	1.68523	3.043742	2.992583	2.914332	2.899724	2.928876	2.92399
2.886752	2.357023	2.1187	2.009975	2.825174	2.777689	2.891366	2.876873	2.762861	2.758252
1.527525	1.247219	2.1187	2.009975	2.925316	2.876147	2.799784	2.78575	2.907806	2.902955

computation. Counting the number of times the computed standard deviation is greater or less than the population value of standard deviation (σ) in each column, we find that for the $(n - 1)$ case, the computed value is less than σ in nine cases and greater than σ in 11 cases. When n is used, 15 computed values are less than σ, and only five are greater than σ. If the table is extended, similar results are obtained. While hardly rigorous, this should be convincing evidence that, indeed, the computation using $(n - 1)$ tends to cluster around σ, while the computation using n tends to cluster around a somewhat smaller value, i.e., the computation using $n - 1$ results in a calculated value of standard deviation that is an unbiased estimator of σ, while using n results in a biased estimator. Why should this be so?

Well, let us consider the expression for standard deviation. The numerator term is the sum of squares of the deviations of the data from the mean value $\bar{X}$. But as with all parametric statistics, the value of standard deviation for the sample is intended to estimate the corresponding value for the population. Because the definition for standard deviation of a population uses n in the denominator when taking sums of squares of differences from the population mean, we can see that a random sample of points will have differences representative of the behavior of the population. In fact, because the size of a population may not be known, a random sample of the population can represent the population, as long as the population mean is used. So, if we compute differences from the population mean, rather than from the sample mean, we expect that these differences will reflect the behavior of the differences that the population members show. We would then use n as the divisor term in the expression.

It can be shown (and we do so in Appendix 4-B) that computing the sum of squared differences from any arbitrary number results in the smallest possible value of this quantity when that arbitrary number equals the sample mean ($\bar{X}$). Hence, in the calculation of standard deviation, the smallest possible value is also attained when the sum of squares is calculated around the sample mean ($\bar{X}$) of the data set. Thus, the sum of squares around the sample mean is smaller than the sum of squares around any possible other value. In particular, the sum of squares around the sample mean is smaller than it would be if sums of squares were computed around the population mean ($\mu = 5.5$ as we have seen in Chapter 3). Also in Chapter 3, we have seen that the sample mean is almost never equal to the population mean. Thus, in the long run, sums of squares computed around a sample mean will be less than the corresponding sums of squares computed around the population mean. But then, if we used this value of sum of squares to compute the standard deviation, we would, in the long run, obtain a smaller value for standard deviation from this computation than from the computation using the population mean. Because we want the sample statistic to tell us the population value for the corresponding parameter, this state of affairs is unsatisfactory. Therefore, we must compensate for this decrease in the sum of squares caused by taking differences from the sample mean rather than the population mean. It is the need to compensate for this decrease in the sum of squares that requires a decrease in the denominator term from n to $n - 1$. That the change needed is exactly that achieved by using $n - 1$ instead of n is beyond what can be shown with the semi-qualitative arguments we are using here; the rigorous derivations are needed to prove that point.

If the decrease in sums of squares is due to the use of the sample mean, then use of the population mean should allow for an unbiased estimation of σ by using n in the denominator. We have provided for testing this in our program. If line 9140 is deleted, and line 9141 is changed to:

$$9141\ \text{XBAR} = 5.5$$

then the program will compute standard deviations around the population mean μ. Table 4-2 demonstrates the results of performing this calculation. Again, for the case of $n = 3$ in Table 4-2 we see that now, when the divisor is n, there are exactly 10 computed values less than the population value, while when the divisor is $n - 1$ there are only six. This confirms our statement that n is the proper value to use when the population mean is available.

Another way to look at this, consistent with our prior discussion of degrees of freedom, is to note that, when the population mean is used, no degrees of freedom are lost by computing the sample mean statistic; therefore, the number of data, n, is indeed the proper number of degrees of freedom to use in this case. On the other hand, if a degree of freedom is used up in computing the sample mean, then this is reflected in the need to use a divisor of $n - 1$, because this is now the number of degrees of freedom left in the data set.

Other important points can be seen in Tables 4-1 and 4-2. The first is that the standard deviation behaves similarly to the mean insofar as:

(1) The sample value of standard deviation (S) almost never exactly equals the population value σ but does tend to be an unbiased estimator of it, when the proper degrees of freedom are used.
(2) The variability of the calculated standard deviation decreases as more and more samples are included in the calculation. To illustrate, we note that when 300 samples are included, the variability is less than when 30 are used, and that is less than when only three are used.

This points up another characteristic of calculations on random data: any statistic has decreased variability when more data is used in its computation.

One last point needs mentioning here: the calculation of standard deviation is a prototype calculation, in that it is the simplest calculation of one standard type of statistical calculation: that is the division of a sum of squares by an appropriate number of degrees of freedom. This calculation occurs over and over again. Indeed, many equations found in statistical texts simply reflect the calculations

TABLE 4-2

Each column contains the standard deviation for n random samples selected from the population of integers 1 – 10. The standard deviation is computed around the population mean, 5.5 rather than around the individual sample means.

$n = 3$		$n = 10$		$n = 30$		$n = 100$		$n = 300$	
$(n-1)$	(n)	$(n-1)$	(n)	$(n-1)$	(n)	$(n-1)$	(n)	$(n-1)$	(n)
2.318405	1.892969	2.592725	2.459675	2.813146	2.765863	3.023477	3.008322	2.799307	2.794638
3.372685	2.753785	3.341656	3.170174	3.002872	2.9524	2.516612	2.503997	2.846696	2.841948
2.715695	2.217356	2.758824	2.617251	3.339936	3.283798	2.862153	2.847806	2.794524	2.789863
4.401704	3.593976	2.953341	2.801785	3.192124	3.138471	2.862153	2.847806	2.817171	2.812472
3.221024	2.629956	3.27448	3.106445	2.330458	2.291288	2.812311	2.798214	2.86076	2.855988
3.221024	2.629956	3.100179	2.941088	2.460025	2.418677	2.893741	2.879236	2.86076	2.855988
4.168333	3.40343	2.915476	2.765863	2.649008	2.604483	2.812311	2.798214	2.870097	2.86531
2.524876	2.061553	3.836955	3.640055	2.738613	2.692582	2.869202	2.85482	2.885206	2.880394
3.657185	2.986079	3.566822	3.383785	2.861757	2.813657	2.728451	2.714774	2.964111	2.959166
3.657185	2.986079	3.17105	3.008322	2.885756	2.837252	2.721036	2.707397	2.802889	2.798214
2.09165	1.707825	3.749074	3.556684	2.921384	2.872281	2.8159	2.801785	2.914042	2.909181
4.046603	3.304038	2.953341	2.801785	3.245687	3.191134	2.938769	2.924039	2.974248	2.969287
3.517812	2.872281	2.758824	2.617251	2.542738	2.5	2.931887	2.91719	2.935767	2.93087
3.657185	2.986079	2.505549	2.376973	3.05975	3.008322	2.879745	2.86531	2.900236	2.895399
4.168333	3.40343	1.900292	1.802776	3.126637	3.074085	2.797907	2.783882	2.882887	2.878078
2.715695	2.217356	2.505549	2.376973	2.403303	2.362908	2.969543	2.954657	2.941458	2.936551
4.730222	3.86221	3.02765	2.872281	2.609664	2.565801	2.721036	2.707397	2.911745	2.906888
3.517812	2.872281	2.013841	1.910497	3.048459	2.997221	2.91461	2.9	2.930065	2.925178
4.513868	3.685557	2.12132	2.012461	2.825378	2.777889	2.945635	2.93087	2.764442	2.759831
2.09165	1.707825	2.321398	2.202272	3.002872	2.9524	2.823065	2.808915	2.908297	2.903446

required to regenerate sums of squares and degrees of freedom from other results, so that they may be combined in different ways. One example is the so-called calculation of "weighted" standard deviations from prior calculations. Inspection of the formulas will reveal that such formulas simply recover the sum of squares in the numerator and the degrees of freedom in the denominator. We point this out in order to keep our discussion at a basic level. As we have discussed previously, division of sums of squares by degrees of freedom is one of the basic statistical calculations, knowing this obviates the need to memorize complex equations that simply perform a specific calculation.

Appendix 4-A

Program to Calculate Standard Deviations of Random Samples

```
10 N=3
15 DIM POP(10), DAT(N)
20 REM randomize timer :rem remove first remark to change
   numbers
30 FOR I=1 TO 10: READ POP(I): NEXT I: REM get data (from
   line 450)
40 FOR I=1 TO 20 : REM repeat 20 times
50 GOSUB 9000 :REM get S.D. of n numbers
300 PRINT S1,S2 :REM print them
350 NEXT I : REM end of loop
400 END : REM end of program
450 DATA 1,2,3,4,5,6,7,8,9,10
9000 REM
9001 REM subroutine to compute s.d.'s from POP
9002 REM
9050 SUMX=0 : REM initialize
9090 FOR J=1 TO N :REM we're going to do n numbers
9100 GOSUB 10000 : REM get a random number
9120 SUMX = SUMX + A :REM generate the sum
9121 DAT(J) = A :REM save data for further computation
9130 NEXT J :REM end of summation
9140 XBAR = SUMX/N : REM sample mean - use 9140 or 9141 but
     not both
9141 REM XBAR = 5.5 : REM population mean
9150 SUMSQUARE = 0 :REM initialize
9160 FOR J = 1 TO N : REM sum of squares
```

```
9161 SUMSQUARE = SUMSQUARE + (DAT(J) - XBAR)^2 : REM sum
     of squares
9163 NEXT J
9170 S1 = (SUMSQUARE/(N-1))^.5 : REM s.d. by (n-1)
     computation
9180 S2 = (SUMSQUARE/N)^.5: REM s.d. by (n) computation
9190 RETURN
10000 REM
10001 REM subroutine to get a random population member
10002 REM
10010 A = RND(A) :REM generate random number
10020 A = INT(10*A) + 1: REM convert to an index
10030 A = POP(A) :REM get a member of the population
10040 RETURN
```

Appendix 4-B

Proof That the Minimum Sum of Squares Is Obtained When Taken Around the Sample Mean

We provide this proof separately, in this appendix, because it requires some math beyond our promise of algebra only.

We begin the proof by defining the sum of squares we wish to evaluate, i.e., the sum of squared differences between n data points, X_i and some arbitrary number X' around which we will take the sums of squares of differences.

$$S = \sum_{i=1}^{n} (X' - X_i)^2 \tag{4-2}$$

We find the minimum by the usual process of taking the derivative

$$\frac{dS}{dX'} = \sum_{i=1}^{n} 2(X' - X_i) \tag{4-3}$$

and setting it equal to zero:

$$\sum_{i=1}^{n} 2(X' - X_i) = 0 \tag{4-4}$$

Separating the summations:

$$\sum_{i=1}^{n} X' = \sum_{i=1}^{n} X_i \tag{4-5}$$

on the right-hand side of Equation 4-5, we multiply by n/n:

$$\sum_{i=1}^{n} X' = n \sum_{i=1}^{n} \frac{X_i}{n} \tag{4-6}$$

But since $\sum_i X_i/n$ equals $\bar{X}$, the equation becomes:

$$\sum_{i=1}^{n} X' = n\bar{X} \tag{4-7}$$

On the left, we note that X' is a constant, and can thus be factored from the summation:

$$X' \sum_{i=1}^{n} 1 = n\bar{X} \tag{4-8}$$

and since the sum of n ones is n, the equation becomes:

$$nX' = n\bar{X} \tag{4-9}$$

or, upon dividing both sides by n:

$$X' = \bar{X} \tag{4-10}$$

Q.E.D.

References

4-1. Mulholland, H., and Jones, C.R., *Fundamentals of Statistics*, Plenum Press, New York, 1968.

5

Introduction to Distributions and Probability Sampling

What do the terms distribution and probability sampling refer to and how does this subject apply to our interest in spectroscopy? We address these questions in this chapter.

If duplicate sample spectra are made on the same aliquot of a (chemical) sample, or on different aliquots from the same sample, one immediately following the first, the two spectra will differ from each other. This situation occurs even when all known precautions are taken to ensure the exact sample presentation geometry. It is not possible for us to predict the exact variation in a spectrum that will occur for any specified aliquot. But when we measure a large number of spectra it becomes possible for us to form a good picture of what the average spectrum of our population of spectra looks like and how much variation we will find in different spectra taken on the same sample. We also find that the larger the number of spectra taken for any individual sample, the more accurate is our determination of the mean and standard deviation of all possible spectra for that sample.

As we have noted previously, all objects sharing a particular characteristic or characteristics in common are collectively termed a population. In dealing with populations of spectra, we would like to know if a single spectrum is really different from another given spectrum (i.e., represents a different sample) or if there is only a chance variation between the spectra and they actually represent the same sample (i.e., are from the same population) with only slight random fluctuations in the spectral pattern. Of course, the purpose of our measurement of a spectrum for any particular sample is to quantify or qualify some characteristic(s) of the sample for which we have an interest.

If the value of our measured characteristic fluctuates, it is called a variable. Variables can be categorized in several ways. Often, particularly when performing quantitative analysis, we distinguish between two types of variables: independent or predictor variables and dependent or response variables. For the purpose of spectroscopic analysis, we can consider independent variables to be those that will ultimately be used to perform analysis (e.g., absorbance or reflectance measurements). As a result of the relationship between the properties of interest and the spectroscopic variables, changes in the independent or predictor variables allow us to determine (read: measure, predict, estimate, analyze) a change in the dependent or response variable (e.g., concentration of analyte, or color). When dealing with spectroscopic analysis, it is important to know if a change in a predictor variable and its effect on a response variable are real, or if a change in response is due to random chance fluctuations in the spectrum. In later chapters we will discuss the meaning and significance of relationships or dependencies found between predictor and response variables.

The exact measurement of a sample spectrum is possible only if all error sources in a measurement are eliminated. We may define error as any deviation of an observed value from a true value. Errors generally come as two major types: systematic and unsystematic (or random) errors. The dictionary defines "systematic" as "made or arranged according to a regular method". Thus, systematic errors are those that are predictable (given enough knowledge) and are often unidirectional, although they may have unpredictable magnitudes. Random errors bring about unpredictable deviation in all directions. Large data sets containing small subsets of samples each with a systematic error component may cause the entire data set to "appear" as if it contained random error.

In Chapters 3 and 4 of this book we have discussed the concepts of the mean and standard deviation for any population. In Chapter 2 we discussed the concepts involved in elementary probability theory. We noted that probability is defined as the fraction of times that a particular event or phenomenon will occur, given that proper conditions are met. A probability (P) equal to 0 indicates that it is impossible for an event to occur, whereas a probability of 1 means that event will occur with absolute certainty. From the laws of probability we could, for example, determine the probability that a given spectrum results from a sample containing a given concentration of analyte, or the probability that two spectra will overlap within a certain tolerance.

We use the concept of distributions to describe the frequency of the occurrence of each specific value, given the entire population of all possible events. As an example, Table 5-1 lists the values of means and standard deviations obtained

TABLE 5-1
List of all possible means and standard deviations obtainable from the population of 10 integers, and the number of times each value occurs.

Means		Standard deviations	
Value	Occurrence	Value	Occurrence
1	1	0	10
1.333333	3	0.5773503	54
1.666667	6	1	48
2	10	1.154701	48
2.333333	15	1.527525	84
2.666667	21	1.732051	42
3	28	2	36
3.333333	36	2.081666	72
3.666667	45	2.309401	36
4	55	2.516611	36
4.333334	63	2.516612	24
4.666667	69	2.645751	60
5	73	2.886751	17
5.333334	75	2.886752	13
5.666667	75	3	24
6	73	3.05505	48
6.333334	69	3.21455	48
6.666667	63	3.464102	24
7	55	3.511885	12
7.333334	45	3.511885	24
7.666667	36	3.605551	36
8	28	3.785939	36
8.333333	21	4	12
8.666667	15	4.041451	24
9	10	4.041452	18
9.333333	6	4.163332	24
9.666667	3	4.358899	24
10	1	4.50925	12
		4.582576	12
		4.618802	12
		4.725816	12
		4.932884	12
		5.196153	6

from all possible combinations of the members of our population of the first 10 integers taken three at a time. Since there are 10 members in the population, there are 1000 possible combinations, and Table 5-1 lists all the means and standard deviations and how often each value of these statistics occurs; there are a total of 1000 of each statistic.

An important point to note here is that, since each value in Table 5-1 represents the mean (or standard deviation) of three members of the original population, taking a random sample from Table 5-1 is identical to taking three random samples from the original population and calculating these statistics. *Therefore, the fraction of times any value occurs in Table 5-1 is the probability of obtaining that value when samples of three are collected from the original population.* Appendix 5-A contains a program written in BASIC which can be used to duplicate this table.

A value for the mean of unity, or a value of 10, each occurs only once [only in the one case where each of the three samples is unity (or 10, as the case may be)]. Since there are 1000 values total, the probability of either of these events is 0.001. A value of zero for standard deviation occurs 10 times (once for each occurrence of the same three integers being chosen). Thus, the probability of its occurrence is 10/1000 = 0.01. Similarly we could calculate the probability of obtaining any value in Table 5-1 from the frequency of its occurrence listed in the table. (We could even predict the frequencies of occurrence. For one simple example, a standard deviation of 4.0 occurs only when the three numbers are four apart. Thus, the triplet 1,5,9 has an S.D of 4.0. This triplet can occur in six ways. The triplet 2,6,10 can also occur in six ways. There are no other triplets possible that meet this criterion, thus there are 12 ways to achieve an S.D. of 4.0.) Now the reason for the results we obtained in Chapter 4 becomes clearer: those results simply represent the actuality of finding these probabilities when we actually did the experiment of taking random samples of size three from the population.

Just as we determined the probability of finding any specific value in Table 5-1, we can also determine the probability of finding a number in any specific range. Thus, it is not possible to obtain a value for the mean that is less than 1 or greater than 10. The probability of either of these events is zero. Five percent of the means lie at or above 8.333..., another 5% lie at or below 2.66... Thus, when means are drawn at random, there is a 90% probability of drawing a mean from the middle two-thirds of the range. We also note, for future reference, that 95% of the means lie between 2.333... and 8.666...

The program in Appendix 5-A is readily modified to produce the corresponding results for all possible combinations of four, five, or more at a time. Even before doing that, however, we can predict certain properties that we will find in the results. Let us consider the results obtained from calculating means and S.D.s of samples of four integers. Again, we will find that a mean of 1 and a mean of 10 will each occur only once. However, since there are now 10,000 values in the table, the probability of obtaining either of these values has dropped to 0.0001. Similarly, we find that, while there are a few more values near the extremes of the range of the population, the vast majority of the newly computed values lie near the middle. Thus, a random sample from the listing of means taken four at a time will show a higher probability of finding a value near the population mean than a random sample of means taken three at a time. This now explains the results of Chapter 4, where we noted that when more samples were included in the calculation of the sample mean, the values obtained clustered more closely around the population mean. This is due simply to the fact that, as more and more values are averaged, it becomes overwhelmingly likely that only values near the population mean will occur, and that values far away from the population mean are so unlikely that they essentially are never found.

The program of Appendix 5-A, modified to compute the statistics from all possible combinations of five samples, confirms these predictions: 90% of the means lie between 3.4 and 7.6 (compared with 2.66... to 8.33... for samples of three) and 90% of the standard deviations lie between 1.414 and 3.9115 (compared with 0.577 to 4.58 for samples of three). The limits corresponding to 95% of the values are 3.0–8.0 for the mean, and 1.1401–4.098 for the standard deviation.

The distribution of a population is the frequency of occurrence of its members, or, in other words, their probabilities as a function of their values. The nature of any distribution is often illustrated graphically by plotting the frequencies versus their values. In the case at hand, we look to Figure 5.1-A as a representation of the results we have obtained with our test population.

This figure shows how the means of the population cluster around the center of the range and tapers out on either end. In general, any set of measurements can be viewed as a subset (or subsample) of a very large (possibly infinite) population; it could even be the entire set of measurements from a smaller population. If random independent samples are drawn from the population, the sample mean ($\bar{X}$) for any sample subset will approximate the population mean (μ). Each sample subset also has a standard deviation (measure of the spread of the data), designated by the symbol "S". We note that the value of S is an estimator of the population standard deviation (which is denoted by σ) when appropriate degrees of freedom are utilized (see Chapter 4).

If the histogram in Figure 5.1-A was set up with a very large (or infinite) number of points (or blocks) on the abscissa (x- or horizontal axis), the edges of the small blocks could be connected and the diagram would look like Figure 5.1-B.

Recalling our histogram in Figure 5.1-A, we point out that the frequency associated with any group, or interval, is given by the area of each block or rectangle. In a continuous sample population (e.g., the concentration of an analyte in many samples of a material) we generally round our variables to the nearest whole unit or to those decimals limited by our instrument or computer. Because there is a continuum of possible values for our variables, we could try to compute the probabilities that our analyte value would equal (say) 10 pi percent analyte. Of course our computed probabilities for these exact analyte concentrations are infinitesimal (read: zero). For our purposes, we are interested

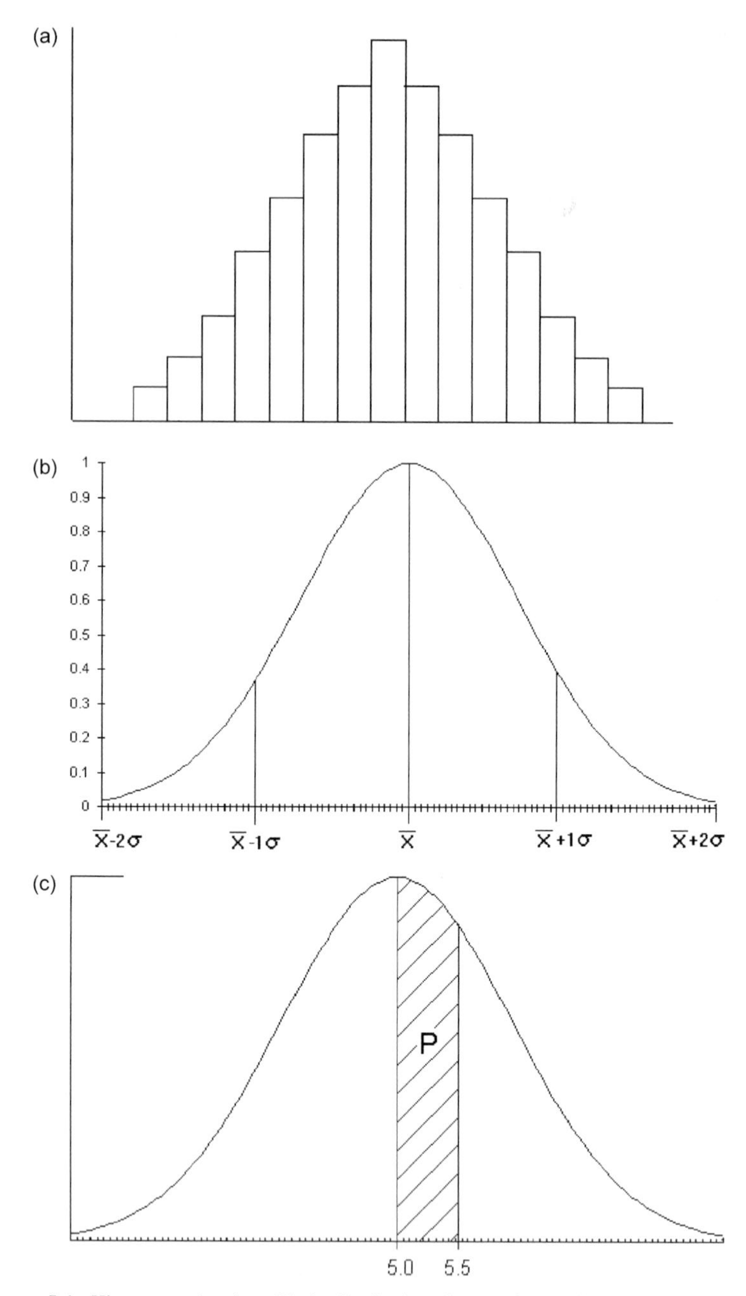

FIGURE 5.1. Histograms showing: (A) the distribution of means from a discrete population; (B) the distribution of means from a continuous population; (C) the probability of obtaining a mean value in a given range within a continuous population.

in the probability of a concentration occurring within a certain interval or within specified limits. For example, we may wish to know how confident we can be that our concentration of analyte is between perhaps 5.00 and 5.10%.

In a method similar to actually counting the values or measuring the areas of the rectangles of our histograms, we may also measure the areas under a continuous curve (Figure 5.1-C). The areas under specified portions of our curve represent the probability of a value of interest lying between the vertical lines enclosing the ends of our area under the curve. Thus, the area between $\bar{X}$ and $+1\sigma$ in Figure 5.1-B represents the probability of selecting a single measurement, object, or event from our population and having the value of the measurement, etc. fall between the mean and $+1$ standard deviation of all possible measurements contained in our population. If Figure 5.1-B is to represent the error associated with the measurement of an analyte concentration, the probability of a single measurement yielding a concentration level within specified limits is given by the area under the curve between those concentration limits (Figure 5.2-B). Thus, depending on the type of distribution function in question, we can compute probabilities for error limits by determining the areas under the curve of that function.

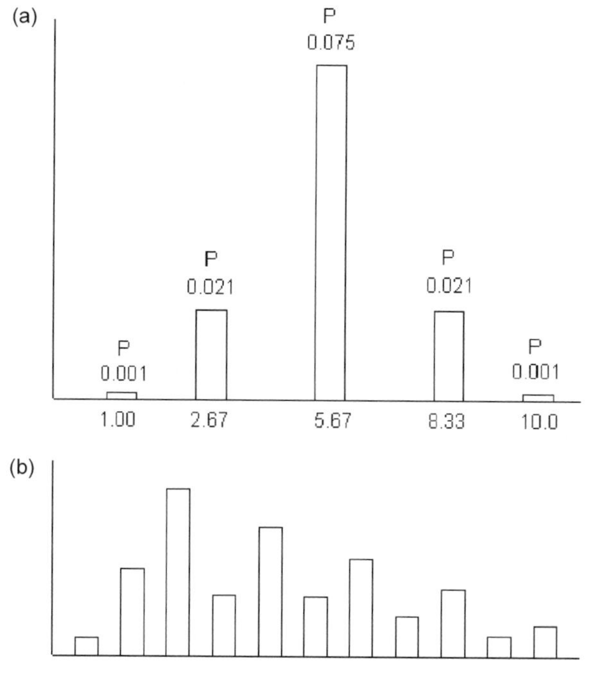

FIGURE 5.2. Portions of the distribution of A: means and B: standard deviations from the (discrete) population of the first 10 integers.

We can (manually, if necessary) plot the frequency of occurrence of various phenomena using histograms to describe the shapes of populations of measurements, events, objects, etc. We can even select small subsets of larger populations and calculate statistical parameters that are assumed to be fairly good estimators of the larger population statistics. Histograms are also usable when describing the possible values of individual statistical parameters. When we draw many subsets from a population and calculate the individual values for a statistical parameter(s) (e.g., mean and variance statistics) we can plot the relative frequency of these computations to arrive at a distribution of the statistic in question. When the distribution of a statistic is defined, we are able to describe the possible range of values for each statistic, and the probability of each value being obtained for a given subset drawn from the overall population.

To illustrate, we look at Figure 5.2-A, which shows a distribution or histogram of the possible means of a given population (Figure 5.1-A). When we know the distribution for a population, we are then able to compute the probability of obtaining a given statistic from a subset of the larger population. The distribution of the statistic cannot tell us *a priori* what a particular subset mean will be, but it does allow us to compute the probability of each event, or "how likely" the occurrence of each possible event will be.

We are getting close to being able to show how to apply statistical concepts to real data, but we are not quite there yet. However, we are ready to start tying together some of the concepts we have been discussing by showing how the concept of a statistical hypothesis test, a tool we have brought up in Chapter 2, can be applied to artificial data, namely, the statistical distributions arising from the population of integers. It works like this: suppose we have collected three pieces of data and calculated their mean. We wish to know if those three data could have come from the population of integers. We begin by noting that, if the data did indeed come from the population of integers, then the sample mean will estimate the population mean (μ). Our null hypothesis is that there is no difference between the mean of our data and the population mean except those that arise from randomly sampling the population. But Table 5-1 lists the means that can arise. Therefore, as we have seen above, values for the mean less than one or greater then 10 are impossible if, in fact, the samples were drawn from this population. If the mean of our data falls outside these limits, then the samples could not have arisen by being picked from the population of the 10 integers we have been working with. In that case, we reject the null hypothesis, and conclude that there is indeed a difference between the means, and that therefore the samples were not drawn from the population of integers. This is the essence of a statistical hypothesis test.

Note that, if the data mean falls within those limits, it does not prove that the data did arise from this population; we can only say that the data is consistent with the null hypothesis. Since a sample mean falling outside the allowable limits *does* prove that the sample data did not arise from the specified population, this explains what we meant, when we stated that hypothesis tests are not symmetrical; one conclusion is much stronger than the other.

When we attempt to describe the distribution of the population of all possible standard deviations of 10 integers drawn at random from a population of the integers (i.e, 1, 2, 3,...,10); we find the histogram shown in Figure 5.2-B. Although the distribution of values is not the same as the distribution of means, we find that these also cluster around the population value, and it is thus possible to perform hypothesis testing on standard deviations also. Thus, using an argument similar to the one above, if a sample of three items has a standard deviation larger than 5.196... that is also proof that the data could not have come from the population of 10 integers, regardless of the mean.

Many types of distributions are described in the literature and involve special considerations. Some of these distributions will be dealt with in future chapters. An excellent discussion of the many distributions that are found in nature can be found in Reference (5-1).

Appendix 5-A

BASIC Program to Generate All Means and Standard Deviations from the Population of Integers 1–10

```
10 N = 3
14 DIM COUNTMEAN(1000),COUNTSD(1000)
15 DIM POP(10),DAT(N),ALLMEAN(1000),ALLSD(1000)
16 MEANTOT = 0:SDTOT = 0: REM number of means and s.d.'s
   saved
20 FOR I = 1 TO 1000: ALLMEAN(I) = 0:ALLSD(I) = 0:NEXT I:
   REM initialize
30 FOR I = 1 TO 10: READ POP(I): NEXT I: REM get data (from
   line 450)
40 FOR I = 1 TO 10 : REM set up all combinations
42 FOR J = 1 TO 10
43 FOR K = 1 TO 10
50 GOSUB 9000 :REM get mean and s.d. of 3 numbers
55 REM PRINT XBAR,S1 :REM diagnostic print
57 REM
58 REM See if XBAR is already in list:
59 REM
60 FOR L = 1 TO MEANTOT: IF XBAR = ALLMEAN(L) THEN GOTO 80
65 NEXT L
```

```
70 COUNTMEAN(MEANTOT + 1) = 1 : ALLMEAN(MEAN
   TOT + 1) = XBAR: REM not in list
75 MEANTOT = MEANTOT + 1 : GOTO 100: REM add and update
   list
80 COUNTMEAN(L) = COUNTMEAN(L) + 1 : REM in list; incre
   ment count
97 REM
98 REM do the same thing for S.D.
99 REM
100 FOR L = 1 TO SDTOT: IF S1 = ALLSD(L) THEN GOTO 150
105 NEXT L
110 COUNTSD(SDTOT + 1) = 1 : ALLSD(SDTOT + 1) = S1
120 SDTOT = SDTOT + 1 : GOTO 350
150 COUNTSD(L) = COUNTSD(L) + 1
350 NEXT K
360 NEXT J
370 NEXT I
390 PRINT MEANTOT,SDTOT : PRINT
400 M = MEANTOT:IF SDTOT > MEANTOT THEN M = SDTOT
410 FOR I = 1 TO M
420 PRINT ALLMEAN(I),COUNTMEAN(I),ALLSD(I),
    COUNTSD(I)
430 NEXT I
435 REM
440 REM sort the list of standard deviations
441 REM using bubble sort due to short list
445 REM
450 FOR I = 1 TO SDTOT-1
460 FOR J = I + 1 TO SDTOT
470 IF ALLSD(I) = < ALLSD (J) THEN GOTO 500
480 TEMP = ALLSD(I):ALLSD(I) = ALLSD(J):ALLSD(J) =
    TEMP
490 TEMP = COUNTSD(I):COUNTSD(I) = COUNTSD(J):COU
    NTSD(J) = TEMP
500 NEXT J : NEXT I
510 REM
520 REM list sorted s.d.'s
530 REM
540 FOR I = 1 TO SDTOT :PRINT ALLSD(I),COUNTSD(I) :
    NEXT I
550 END
4500 DATA 1,2,3,4,5,6,7,8,9,10
```

```
9000 REM
9001 REM compute means and s.d.'s from the population in
     array DAT
9002 REM
9050 SUMX = POP(I) + POP(J) + POP(K)
9060 DAT(1) = POP(I):DAT(2) = POP(J):DAT(3) = POP(K)
9140 XBAR = SUMX/N : REM sample mean- use 9140 or 9141 but
     not both
9141 REM XBAR = 5.5 : REM population mean
9150 SUMSQUARE = 0 :REM initialize
9160 FOR M = 1 TO N : REM sum of squares
9161 SUMSQUARE = SUMSQUARE + (DAT(M)-XBAR)^2 : REM sum
     of squares
9163 NEXT M
9170 S1 = (SUMSQUARE/(N-1))^.5 : REM s.d. by (n-1)
     computation
9180 REM S2 = (SUMSQUARE/N)^.5 : REM s.d. by (n) compu
     tation - CANCEL
9190 RETURN
```

References

5-1. Johnson, N.L., and Kotz, S., *Continuous Univariate Distributions*, 1st ed., Houghton Mifflin Company, Boston, 1970.

6

The Normal Distribution

The mathematical expression:

$$Y = \frac{1}{\sigma(2\pi)^{1/2}} \exp\left[-\frac{1}{2}\left(\frac{X-\mu}{\sigma}\right)^2\right] \tag{6-1}$$

is used to describe a model for the normal or Gaussian distribution. The shape of this distribution is shown by both Figures 5.1 and 5.2. In the absence of any other information, the Normal distribution is assumed to apply whenever repetitive measurements are made on a sample, or a similar measurement is made on different samples. The reason for making this assumption will be discussed in a forthcoming chapter.

Properties of the Normal Distribution

Two very important characteristics of this distribution, which distinguish it from the distribution of the averages of the 10 integers that we have been working with, must be noted. The first characteristic is that the normal distribution is continuous. The distribution of the averages of the integers 1–10 was discrete. The discrete nature of that distribution allowed us to determine the probability of obtaining a particular value from the distribution by actually counting the number of averages at any particular value. We noted that in the above case we could also enumerate the number of averages falling within any particular range.

In the case of a continuous distribution we cannot count entries, because there are infinitely many of them; nor can we determine the probability of finding any particular exact value, because the probability of obtaining any one exact value is zero. Thus, in the case of the normal distribution, or indeed any continuous distribution, we can determine probabilities only by specifying a range of values that a selected number may fall into.

The second key characteristic of the normal distribution that differs from the distribution of the average of the 10 integers is that it is infinite in extent. This difference is crucial when we wish to perform a hypothesis test: when the extent of the distribution is infinite we cannot set an absolute limit to the values that distinguish candidates for membership in the specified population from those that cannot be members because they lie outside the range of possible values.

Here is where the probabilistic considerations come in.

Sixty-eight percent of the area of the normal distribution curve lies within 1 standard deviation of the center value (the mean). Ninety-five percent of the area lies within 1.96 (let us say 2) standard deviations and 99% lies within approximately 2.5 standard deviations of the mean. This means that if numbers (read: data) are drawn at random (read: measured) from a set that follows a normal distribution, then 68% of those numbers will be within one standard deviation of the mean, 95% will be within two standard deviations of the mean and 99% will be within 2.5 standard deviations of the mean. Recalling our definition of probability, this is the same as saying that the *probability* of finding a number within one standard deviation is 0.68, within two standard deviations is 0.95 and within 2.5 standard deviations is 0.99.

So what?

So this: pick a limit (let us say 2.5 standard deviations for the sake of discussion). Any value (datum) must either fall within the limits or outside the limits: these are mutually exclusive outcomes. By Equation 2-6 of Chapter 2, $P + \sim P = 1$. Thus, since the probability of a value falling within the ± 2.5 standard-deviation limit is 0.99, the probability of a value from this distribution falling outside this limit is 0.01.

This is the key to performing hypothesis tests on real data: we cannot set an absolute cutoff as to where we can expect to find no values from the population against which we are testing data, but we can set a limit beyond which we consider it *very unlikely* to find a member of the population. If we collect data and find that it is beyond the probabilistic limit, then we can say that we have evidence that our data did not come from the test population, because it is too unlikely that a sample from the test population would have been found outside the limit.

This concept is general: we can even apply it to the distribution of averages from the set of integers. We have noted that 95% of the means of samples of three integers drawn at random lie between 2.333... and 8.666... (see Chapter 5).

Initially we noted that if the mean of a sample of three items fell outside the range 1–10 we would know that the items were not taken from the population of the 10 integers. Now we can tighten the limits to 2.333... to 8.666... because it is "too unlikely" that an average outside that range would have been obtained if, in fact, the data belonged to that population.

That brings us to ask two key questions:

(1) How do we know what values of various statistics correspond to the various probability levels of interest?
(2) How unlikely is "too unlikely"; i.e., how do we set the cutoff limit to perform hypothesis testing against?

The answer to the first question is actually very straightforward. Every statistic has a distribution of values. Consequently, every value of that statistic corresponds to a well-defined value of the twin probabilities that a random selection from the population following that distribution will be greater/less than that value. It remains only to determine the probability levels corresponding to the values of the appropriate distribution; there are several methods for doing this:

(A) If the distribution is well defined mathematically, it can be integrated over the range of interest. If the distribution is sufficiently well behaved, the integration could be done analytically. In general, distributions of statistical interest are not that well behaved; integration of these distributions must be done using numerical methods. It is well known (for example) that the integral of the Normal distribution cannot be determined analytically (except in special cases), yet it is one of the simplest distributions we will be involved with.

(B) An exhaustive compilation of the possible values of the distribution will allow direct evaluation by enumeration of the probability levels corresponding to any value. This is the approach we took to evaluating the probability levels of the means of samples of three from the population of the integers 1–10. This approach is suited only to discrete, and relatively small, populations.

(C) A related method is to use a Monte Carlo technique to generate a large number of random samples from the distribution of interest. While the compilation of values will not be exhaustive in this case, enumerating the fractions falling within (or outside) various ranges will allow determination of the probability levels as in case B. This approach has recently been put to use in evaluating a wavelength-selection method for near-infrared reflectance analysis (6-1).

(D) We saved the best for last. In practice, none of the above methods are used on a routine basis. For all the common distributions, and even many uncommon ones, the probability levels corresponding to values of interest have been calculated (using one of the above methods) checked to high accuracy, and have been set down once and for all in tables. To find the probability level for a statistic of interest, one need only consult a table for the distribution of that statistic. Short tables of the common statistics are included in virtually every book dealing with statistics. More exhaustive tables, and tables of the less common statistics are found in books devoted exclusively to their presentation (see, e.g., References (6-2,6-3)). We strongly urge readers, especially those who expect statistics to be useful to them, to obtain some compilation of statistical tables, as this is the only sure way to have accurate values available.

The use of the tables is straightforward. The tables list the probability of a result being between the lower limit for the statistic in question, and the given value. To take a simplified example, we know that for the Normal distribution, the probability is 68% that a value will lie between −1 S.D. and +1 S.D. Therefore, 32% lies outside those limits, 16% on each side. Therefore, the table for the Normal distribution has 0.158 listed for the −1 S.D. value, 0.50 for the 0 S.D. value and 0.841 for the +1 S.D. value. For symmetric distributions, of which the Normal distribution is one example, sometimes the central area, i.e., the probability of a value lying within a given distance of the mean, is listed instead. Values from one type of table can be converted to the equivalent value in the other type, as we did above, simply from the twin facts that area outside the value of interest is divided equally, and that the mean falls at exactly the 50% point.

To answer the second question: "how do we set the probability level for the limit at which we will reject a null hypothesis?" is more difficult. A full analysis of the situation involves determining the cost (and it is indeed often *monetary* cost!) of incorrectly rejecting the null hypothesis versus the cost of incorrectly accepting the null hypothesis. We do not have space here to discuss the subject in the detail it deserves, but readers looking for more information on this topic can find it in practically any book on elementary statistics; it is often labeled with a topic title something akin to "Producer's versus Consumer's Risk". At this point we will simply note that the standard test levels for performing hypothesis testing in the absence of a full analysis are 95 and 99%, and we will justify these values by pointing out that someone who can prove mathematically that he is right 95% of the time is doing very well indeed!

Nomenclature for Hypothesis Testing

At this point it is appropriate to define some more nomenclature, and remind ourselves of a caveat concerning proper use of hypothesis tests.

As we have noted several times by now, every population has a set of parameters that describe its characteristics. Selecting a random sample from that population and calculating the statistics corresponding to the population parameters gives a number for each statistic that is an *estimate* of the parameter.

Such estimates are called *point estimates* because a single calculation results in an estimate of the parameter at only one point on the distribution of the statistic.

The distributions of statistical interest are defined mathematically as extending to infinity. As we have noted, in order to use these distributions (as we must, because they describe the probabilities of obtaining particular values of various statistics) we must demarcate finite regions of their domains, and determine the probability of obtaining a value within the given range. The range of values that is marked off, and that we will perform hypothesis testing against, is called the *confidence interval*. The probability of obtaining a value within the demarcated range is called the *confidence level*. Thus, in the Normal distribution, the confidence interval from -2 to $+2$ standard deviations corresponds to a confidence level of 95%. In the distribution of averages of three integers, the confidence level of 95% corresponds to a confidence interval of 2.333... to 8.666... and the 95% confidence level of standard deviations from that population corresponds to a confidence interval of 0.577–4.58 (see the tables in Chapter 4 and review the discussion about them).

The ends of the range, defining the limits beyond which we will decide that a given statistic causes us to reject the null hypothesis, are called the *critical values* of the distribution. Clearly, the confidence interval and the critical values depend upon the confidence level at which we perform a given hypothesis test.

The *expected value* of a statistic is another term we will want to use. There is a rigorous mathematical definition for this term, which is based on integrating the product of the values of the statistic and their respective probabilities. However, we will note it here as what we have seen previously: the expected value of a statistic is the value that it tends to cluster around when the statistic is computed many times. As we have seen, the expected value is not necessarily the population parameter, if the statistic is biased.

Now we come to our caveat. The probability levels used to compute the values in the tables are correct only if the hypothesis test is done correctly. In particular, these probabilities are based upon the assumptions that only a single experiment is performed, and that the null hypothesis is correct; then the value obtained for a statistic is due only to the random variations of the phenomenon under test. These assumptions are what allow us to reject the null hypothesis if the value obtained from the experiment lies beyond the critical values. However, we must ensure that we have not violated the other assumptions upon which the hypothesis test is based. If more than one experiment is performed (or a statistic is calculated more than once), then we must apply the considerations we have discussed in Chapter 2, and adjust the critical levels of the hypothesis test appropriately. One place where this requirement is commonly violated is during computerized searches for the optimum wavelengths to be used for quantitative spectroscopic analysis, an exercise that is becoming more and more common. Similarly, we must also check that we are in fact using the correct distribution for the data at hand.

The characteristics of distributions, in particular the Normal distribution, can be useful, when applied appropriately. We will now discuss how probability theory and the properties of the normal distribution can be applied to a situation of spectroscopic interest. (We finally get to work with some real data!)

Statistics and the 100% Line

Figure 6.1-A shows what most mid-infrared spectroscopists have surely seen: the 100% line of an FTIR spectrometer, between 500 and 4000 wavenumbers. This spectrum was obtained, as such spectra are usually obtained, by measuring the sample and reference path single-beam spectra with no sample present, and dividing one by the other in the computer that is invariably associated with the FTIR instrument. The 100% line is often used as a measure of the noise level of the instrument. The one presented in Figure 6.1 was originally measured at two wavenumber resolution, but only every fifth point is plotted in Figure 6.1-A.

At the recent FACSS meeting (this article was originally written in December 1986, so the FACSS meeting was only a few months ago), we conducted a short survey of methods used to calculate the noise level by different manufacturers. Some items were common to all manufacturers: invariably the data used was taken from the spectral region between 2000 and 2200 wavenumbers. Spectroscopically this is a desirable region to work in because it avoids effects caused by normal atmospheric constituents, which might affect the results. Of course all marketing departments like it because it shows the instruments off to best advantage by avoiding regions of excess noise, such as the region around 500 wavenumbers, where energy limitations cause the apparent noise level to increase.

From the statistical point of view, we give this procedure a mixed review. On the negative side, calculating over a prespecified range of wavelengths does not provide an unbiased estimate of the overall noise of the 100% line. Indeed this wavelength range used was chosen partly because it results in the lowest noise value. On the positive side, as Griffiths shows (6-4) the instrument noise really shows up in the individual single-beam spectra obtained after transforming the interferogram; this noise we expect *a priori* to be normally distributed. When we calculate a 100% line, we use a division, which is a non-linear process. Our *a priori* expectation, therefore, is that the noise of the 100% line will *not* be normally distributed but will have a distribution skewed toward higher values, and will approximate a normal distribution only where the noise level is low. For this reason, avoiding regions of high noise is desirable for calculations that depend upon the distribution of the noise. Another positive aspect is that all manufacturers use the same spectral region, so that in a practical sense it allows inter-instrument comparisons, even though the true noise values are not obtained.

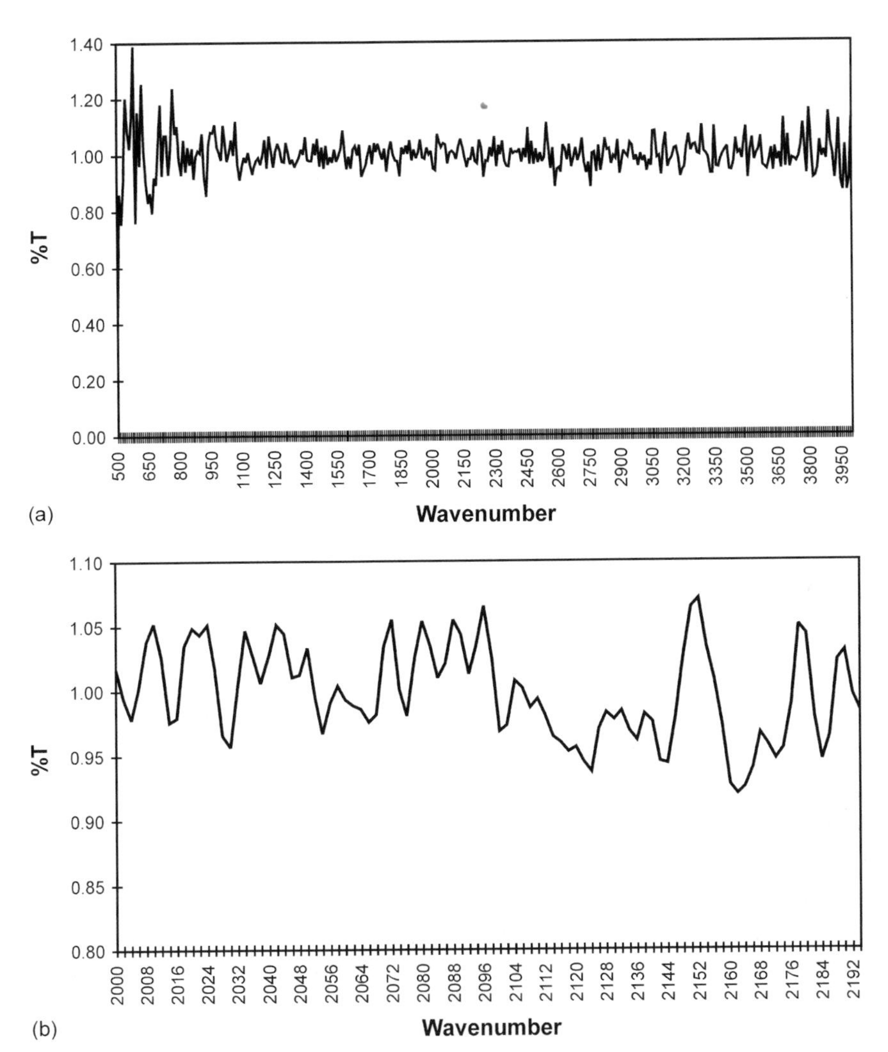

Figure 6.1. 100% line from an FTIR mid-infrared spectrometer, expanded to show the noise level. Note that the instrument performance has been deliberately degraded to increase the noise level. Figure 6.1-A shows the full spectral range, plotted at 10 wavenumber intervals. Figure 6.1-B shows the range 2000–2196 wavenumbers, plotted every 2 wavenumbers; corresponding to the data listed in Table 6-1.

A second reason to avoid combining regions of high and low noise level is even more fundamental. A statistic, as we have noted previously, is intended to inform us about the corresponding parameter. When such large variations of the standard deviation are seen, we have reason to suspect that the *parameter* σ of the different regions of the spectrum is not the same. This leads us to point out a very important rule of statistical calculation: *combining statistics that estimate different parameters is invalid.* The reason for this rule should be immediately obvious from the relationship of parameters and statistics: since a statistic is supposed to estimate a parameter, combining statistics corresponding to different parameters gives us a number that does not estimate *anything* in particular. The implication of this is that there is no single unbiased descriptor of the noise level of the 100% line if, in fact, the parameters corresponding to the noise differ at different wavelengths.

We have not yet discussed methods of testing the question of whether two statistics estimate the same parameter. In addition, examining Figure 6.1 gives us no reason to suspect that more than one value of (sigma) need be invoked to describe the noise level in the region around 2000 wavenumbers. All in all we tentatively conclude that using this limited spectral region is satisfactory because the noise level is used to compare different instruments and all instruments are measured in the same spectral region. Hopefully, the measurements made will all be equivalent, and thus directly comparable. Let us examine this.

In our survey, we found that, while the noise of most instruments was reported as RMS (i.e., standard deviation) actually a variety of methods are used to compute the noise of the 100% line. Some manufacturers do actually calculate according to the formulas we have presented in Chapter 4. One manufacturer fits a straight line to data in the 2000–2200 wavenumber range (to avoid inflating the RMS value by possible tilt of the 100% line) and calculates the RMS difference of the individual points from the line. We will examine this procedure further in the next chapter. By far the greatest number of manufacturers measure the average-to-peak, or the peak-to-peak difference (i.e., the range), and divide by a factor (of 2.5 or 5, respectively) to convert to RMS. One manufacturer reverses this procedure, computing RMS and multiplying by 5 to determine the peak-to-peak difference. In all cases the actual wavelength range, or number of data points used, is chosen arbitrarily. When we asked how many points to use, the response was usually: "what looks good" or "use enough points to be statistically meaningful"(!).

Well, how many points should we use? There is a danger in doing this type of calculation, which we take this opportunity to warn against. A temptation often exists, especially when uncommon statistics are needed, for which no tables are readily available, to simplify the situation and get a result that "we hope will be about right". We will go through an exercise of this sort for the case at hand to illustrate the danger of such an approach.

The following line of reasoning is appealing: From the properties of the Normal distribution we know that 99% of the data lie within ± 2.5 (actually 2.58) standard deviations of the central value. Thus, if we measure 100 data points, 99 should be within these limits, and one should fall outside them. Not very far outside, but outside. In fact, we should expect it to be just barely outside.

Further than that would be too unlikely; only one point out of 200 should be beyond 2.82 S.D.s, for example, and one in 150 beyond 2.71 S.D.s, rather than 1 in 100. Suppose we collect the data points one at a time and watch how they fall, what would we expect to see? Well, the region of greatest probability is around the mean, so the first few points should fall close to the mean. As we collect more and more points, some of them will start to fall at some distance from the mean. As we continue to collect more data, they will spread out more and more, and as the number of data collected approaches 100, the data points will start to crowd out toward the 99% limits on both sides. The 100th point will then be crowded just over the limit.

Of course this is the expected behavior, not necessarily the actual behavior: the probability that it is actually the 100th point that will be the one beyond the limit is also only 1%. But in the long run, by the time we have collected 100 data points, they have spread toward the 1% limits on both sides, i.e., the range has spread to ± 2.58 standard deviations, with the 100th point just going over. From this we would conclude that the expected range of the data is the distance between the limits at which, for the total number of data points, the fraction of points outside the limits equals the fraction of the data represented by one point. Thus, we expect that 100 points will spread out over 5.16 standard deviations.

Similarly, 5%, or 5 out of the 100 data points would be expected to be beyond ± 1.96 standard deviation. But this same 5% would be 1 data point out of 20. So if we collected 20 data points, by arguing similarly to the case of 100 we come to the conclusion that we expect these 20 data would have a range of 3.92 standard deviations. By other similar arguments we could calculate the range expected to correspond to any given number of data points.

So how many data points must we measure in order for the expected range to be exactly 5 standard deviations? Consulting a table for the normal distribution (e.g., Reference (6-4), p. 8) we find that 98.758% of the data falls within exactly ± 2.5 S.D.s of the mean, leaving 1.242% outside those limits. From that we readily calculate that we should use 80.51 data points. Since we cannot collect a fraction of a datum, we conclude that we must measure 81 data points in order for the expected value of the range to be five times the standard deviation.

This determination of the number of data needed, derived from the table of normal distributions, looks reasonable and seems to follow the known laws of probability. Unfortunately it is wrong, and herein lies the danger. Readers knowledgeable in statistics will have already have consulted a table and found the discrepancy. The discrepancy is caused by the fact that the calculations as we did them were based on a simplistic consideration of the probabilities involved; we did not take into account the exact probabilities of finding the extreme data point at various distances outside the limit, or the probability of finding the next-to-last point a given distance inside the limit, or the probability that even the most extreme point will not be beyond the limit, or even the probability of having two points beyond the limit (which admittedly is small, but still non-zero). After all, if we had 100 data points, we expect that only four of them will lie between 1.96 and 2.58 standard deviation (the 1 and 5% limits). If you think about it you realize that it is not really all that likely that one of these four will be exactly at the limit. We deliberately led you down the garden path as a warning to not trust intuition or seemingly reasonable arguments. Only the rigorously evaluated results are correct.

How far are we off? The range expected for random samples from a Normal distribution is also a statistic that has been calculated by integrating the exact mathematical expressions, taking all the necessary probabilities into account, to high accuracy, and compiling the results into tables. From the tables (see, e.g., Reference (6-2), p. 140) we conclude that we must actually measure exactly 98 data points in order for the factor of 5 to be a correct conversion between RMS and the range of the noise in the data. Note: it is the number of data that is important, not the spectral range that the data cover. For the data here, with 2 wavenumber resolution, 98 data points just about cover the 2000–2200 wavenumber section, but at other resolutions, this number of data will occupy different wavelength ranges.

More than 98 data points will, in the long run, tend to give values for the range that is too high, and fewer points will tend to give values that are too low. Conversely, calculating RMS from the range will result in values for the RMS that are too high if fewer points are used, and too low if more points are used.

Let us apply these considerations to the data of Figure 6.1. The list of 98 transmittances corresponding to the 100% line between 2000 and 2194 (inclusive) is in Table 6-1. From these data, we find that the mean is 0.998, and the standard deviation is 0.03664. The range is the difference between

TABLE 6-1
Data from 2000 to 2194 wavenumbers at 2 wavenumber intervals (98 data points total).

1.01675	0.99426	0.97815	1.00185	1.03796	1.05197	1.02729	0.97602
0.97905	1.03500	1.04869	1.04373	1.05123	1.01660	0.96573	0.95683
1.00318	1.04686	1.02642	1.00580	1.02712	1.05115	1.04444	1.01078
1.01233	1.03355	0.99565	0.96700	0.99109	1.00390	0.99333	0.98859
0.98587	0.97594	0.98174	1.03423	1.05513	1.00095	0.98053	1.02544
1.05399	1.03481	1.00984	1.02067	1.05475	1.04368	1.01268	1.03551
1.06541	1.02485	0.96876	0.97317	1.00750	1.00194	0.98689	0.99359
0.98108	0.96458	0.96010	0.95280	0.95618	0.94528	0.93728	0.97044
0.98307	0.97744	0.98416	0.96874	0.96158	0.98189	0.97568	0.9452
0.94344	0.98005	1.02562	1.06448	1.07073	1.03467	1.00780	0.97361
0.92730	0.91986	0.92542	0.94040	0.96698	0.95856	0.94697	0.95498
0.98902	1.05032	1.04335	0.97989	0.94635	0.96461	1.02341	1.03047
0.99650	0.98378						

the high and low values: $1.07073 - 0.91986 = 0.15087$. Thus, the range is $0.1508/0.0366 = 4.12$ times the standard deviation.

Well, that is not exactly 5, but then the range is a statistic, and will show variability, just like any other statistic. The real question is: is 4.12 times the standard deviation beyond the limits of likely values, given that we have used 98 data points? We now know how to answer this question: we will do a hypothesis test. The null hypothesis (Ho) is that 4.12 is the same as 5.0 except for random variability. The alternative hypothesis (Ha) is that the range is not close enough to 5.0 for the difference to be attributed to random chance. There is a formalism associated with hypothesis testing: formally, it is written this way:

$$\text{Ho}: \text{ Range} = 5.0$$

$$\text{Ha}: \text{ Range} \neq 5.0$$

The table we need to consult is the distribution of the range for samples of 98. The closest the readily available table (Reference 6-2, p. 139) comes is for samples of 100. Examining this table we find that the critical value for this statistic changes very slowly at that point, so it is safe to use the critical values for samples of 100. We find that the critical values at the 95% confidence level for samples of 100 are:

$$3.965 < \text{range} < 6.333$$

The value we found, 4.12, is comfortably within those limits; we conclude that the fact that the value is not exactly 5 can be accounted for by chance random variability alone.

Acknowledgements

The authors thank Dr. David Honigs for supplying the data used in Figure 6.1 and Table 6-1.

References

6-1. Montalvo, J., Paper #242, Rocky Mountain Conference, Denver, CO, Aug. 3–7, 1986 (pub).
6-2. Owen, D.B., *Handbook of Statistical Tables*, Addison-Wesley, Reading, MA, 1962.
6-3. Pearson, E.S., and Hartley, H.O., *Biometrika Tables for Statisticians*, Biometrika Trust, University College, London, 1976.
6-4. Griffiths, P., and de Haseth, J.A., *Fourier Transform Infrared Spectroscopy*, Wiley, New York, 1986.

7

Alternative Ways to Calculate Standard Deviation

In this chapter we digress from the development of statistical theory to discuss a sidetrack that is both interesting and useful. In Chapter 4 of this book we presented the formulas for calculating standard deviation for populations and samples. We now apply the formula to particular cases.

Alternative Method #1

We start with the formula for calculating the standard deviation of a set of samples (recalling that Roman letters are used to label sample statistics):

$$S = \sqrt{\frac{\sum_{i=1}^{n}(X_i - \bar{X})^2}{n-1}} \tag{7-1}$$

Let us evaluate this expression for the case of a sample of size two:

$$S = \sqrt{\frac{(X_1 - \bar{X})^2 + (X_2 - \bar{X})^2}{2-1}} \tag{7-2}$$

We also note that, in the particular case of two samples, $X_1 - \bar{X} = X_2 - \bar{X}$. We can therefore rewrite Equation 7-2 as:

$$S = \sqrt{2(X_i - \bar{X})^2} \tag{7-3}$$

Note that the denominator term has disappeared, because it has become a division by unity. Again, in the special case of two samples, $X_1 - \bar{X} = (X_1 - X_2)/2$. Making this substitution in Equation 7-3 yields:

$$S = \sqrt{2\left(\frac{X_1 - X_2}{2}\right)^2} \tag{7-4}$$

Finally, letting D equal the difference $X_1 - X_2$, we find:

$$S = \frac{D}{\sqrt{2}} \tag{7-5}$$

Equation 7-5 estimates the population standard deviation, σ, with one degree of freedom. It is a rather poor estimate because, as we shall find when we get around to discussing the distribution of standard deviations, the confidence interval for standard deviation with only one degree of freedom is very wide indeed.

Nevertheless, S^2, as calculated from Equation 7-5, represents an *unbiased* estimate of σ^2, i.e., the expected value of variance calculated from the expression in Equation 7-5 equals σ^2.

If Equation 7-5 is such a poor estimator, then what good is it? The answer to this question comes from the fact that, if we have more than one estimate of a parameter, we can combine these estimates to give a single estimate that is better than any of the individual estimates (contrast this with our warning against combining statistics that estimate *different* parameters). This process is known formally as *pooling*. In the case at hand, we can pool many poor estimates of σ, obtained by applying Equation 7-5, in order to obtain a good estimate of σ with many degrees of freedom. Any estimate of σ with a given number of degrees of freedom is exactly as good as any other estimate of σ with the same number of degrees of freedom, regardless of whether the estimate is obtained from pooling estimates as done here, or by direct calculation, as we have discussed previously. There is at least one situation in which this process is particularly useful.

When spectroscopy is used to perform quantitative analysis, the question of the error of the analytical method is always very important. Often a multivariate mathematical/statistical technique, such as multiple regression, principal component analysis, or partial least squares is used to develop a calibration equation. We are getting a bit ahead of ourselves here, because we do not plan to discuss these topics in depth yet; but one aspect of the calibration process is pertinent.

The error of the spectroscopic analytical method is usually taken as the RMS (root mean square) difference between the instrumental/spectroscopic results and the "known" or reference values for the materials comprising the calibration, or "training" set. Alternatively, the RMS difference (note that RMS difference is the chemist's and engineer's term for standard deviation) is calculated for the agreement (or rather, the lack thereof) between the instrument and a set of "unknowns", not in the training set. However, this calculation is actually the calculation of the *combined* error of the instrumental/spectroscopic method, and the error of the method used to obtain the reference values for either the training set or the test set of "unknowns". This is particularly true when the materials are not synthetic mixtures that can be made up gravimetrically, or by a method of similar high accuracy, but natural products, or real specimens (7-1) that must be measured "as is". We often encounter this situation in near-IR analysis; from the literature and talks at recent meetings, mid-IR analysis is also becoming involved with similar types of specimens.

It is therefore useful to be able to measure the accuracy of the instrument and the reference method independently, in order to tell where the error lies. Since we are spectroscopists, we will put all the onus on the reference method, and discuss ways to measure the accuracy of the reference method against which we will calibrate our instruments. In principle, this could be done easily: simply split a specimen into many aliquots, have each aliquot measured by the reference method for the constituent we wish to calibrate, and apply Equation 7-1 to the data.

In practice, things are not so simple (are they ever?). It is well known in analytic practice that the reference methods, particularly when wet chemistry is used, are themselves subject to a variety of error sources. These include not only the errors of reading burettes, etc., but also include errors due to using fresh versus old reagent, errors due to the analyst, etc. A good discussion dealing with these matters from a statistical point of view can be found in Reference (7-2).

In addition to not necessarily including all these error sources, which represent the *true* error of the reference method, there are two major problems with blindly applying Equation 7-1 to a set of aliquots of a specimen. The first problem is that, while we would have measured the error of the reference method (actually we have only "estimated" the error; this procedure would represent only one sample of the population of errors of that method, where we use the word "sample" in its strict statistical meaning), we still would not have reference values for the set of specimens we wish to use for our training set. Furthermore, when we sent the training set of specimens for analysis, we would have no guarantee that the errors in the analyses of the training set would come from the same population as the population whose errors we measured.

The second major problem is that many methods are known to have errors whose magnitudes depend upon the amount of the constituent being measured, or upon some other property of the specimens. In such cases, there is a chance that the estimate we obtain for the error is not representative of the value for the population of specimens, but only for the one specimen that we used.

The way around these problems is to have each of the training set specimens analyzed in duplicate. Preferably, the duplicates should be "blind", i.e., the analyst should not know which of the sets he is analyzing are aliquots of the same specimen; also, an attempt should be made to spread the duplicates out (if possible) over different times, analysts, etc., in order to get an estimate of the true total error of the reference method (again see Reference (7-2)). This procedure will then allow us to obtain both a reference value for each specimen in the training set, and an estimate of the error of the reference method on the actual set of specimens that will be used to calibrate the instrument. This is so because every measurement we make contains two components: the true value of the analyte (i.e., a population value; what we would obtain if we could measure infinite number of aliquots from that specimen), plus a contribution from the error of the method. The implicit assumption is made that the error of the method enters independent of the true value of the analyte. Then the estimate of error for each specimen is an estimate of the same σ, and they can be combined.

Having obtained the paired data for each specimen, the calculation of the error of the reference method is straightforward. We wish to pool the data from each of the specimens. As we have mentioned previously, the standard deviation is the square root of the sum of squares divided by the number of degrees of freedom. This is true for both the individual and pooled standard deviations. Thus, the sum of squares in Equation 7-5 is $D^2/2$. The number of degrees of freedom in Equation 7-5 is one. This is the 1 that disappeared between Equations 7-2 and 7-3. In order to pool the results, we separately add together the sums of squares, and the degrees of freedom for the specimens in the training set, to get the total sum of squares and the total degrees of freedom:

$$\text{Total sum of squares} = \sum_{i=1}^{m} \frac{D^2}{2} \tag{7-6A}$$

$$\text{Total degrees of freedom} = \sum_{i=1}^{m} (1) = m \tag{7-6B}$$

where m now represents the number of individual specimens that are used.

Combining Equations (7-6A) and (7-6B) we get the expression for the pooled standard deviation of the reference method:

$$\text{S.D.} = \sqrt{\frac{\text{Total sum of squares}}{\text{Total degrees of freedom}}} = \sqrt{\frac{\sum_{i=1}^{m} D^2}{2m}} \tag{7-7}$$

Note that defining standard deviation as the total sum of squares divided by the total degrees of freedom is general: we need not limit ourselves to paired data; doing so leads to a convenient computational expression, but we are not limited to that case. The numerator term under the radical of Equation 7-1 is the sum of squares for any number of readings, similarly the denominator $(n - 1)$ represents the number of degrees of freedom. The values for sums of squares and degrees of freedom can be added up over different specimens regardless of how many measurements were made for each specimen. Indeed, there is even no requirement that the number of readings be the same for each specimen; each one will make its appropriate contribution to the pooled value when the totals are computed.

Alternative Method #2

A second alternative method of computing standard deviation is an outgrowth of the first. In fact, while the usage of pooled standard deviations for the purpose described above is valid, useful and self-sufficient, another reason we developed that derivation was so that we could proceed still further. A good example of where this further development might be used was discussed in Chapter 6. In our short survey we found that one manufacturer of FTIR spectrometers measures the noise level of the 100% line of his instrument by fitting a straight line to the data in the 2000–2200 wavenumber region, summing the squares of differences between the data points and the line, dividing by $n - 1$ and taking the square root. The purpose of computing the noise this way was to avoid inflation of the noise figure by possible non-flatness of the 100% line; calculating the S.D. according to Equation 7-1 would, indeed, include the variation of the data due to tilt of the 100% line in the noise figure. Aside from the relatively minor point that the divisor should be $n - 2$ rather than $n - 1$ (since two statistics: a slope and an intercept are calculated, therefore two degrees of freedom are lost) this is a reasonable approach to avoiding inflation of the noise figure from that source—assuming that the deviations, if any, are linear.

Offhand, there would seem to be no justification for such an assumption. If the 100% line is not flat, how can we assume that the deviations have any particular form?

Fortunately, there is a way to deal with the situation that does not require making any assumptions about the nature of the departure of the 100% line from

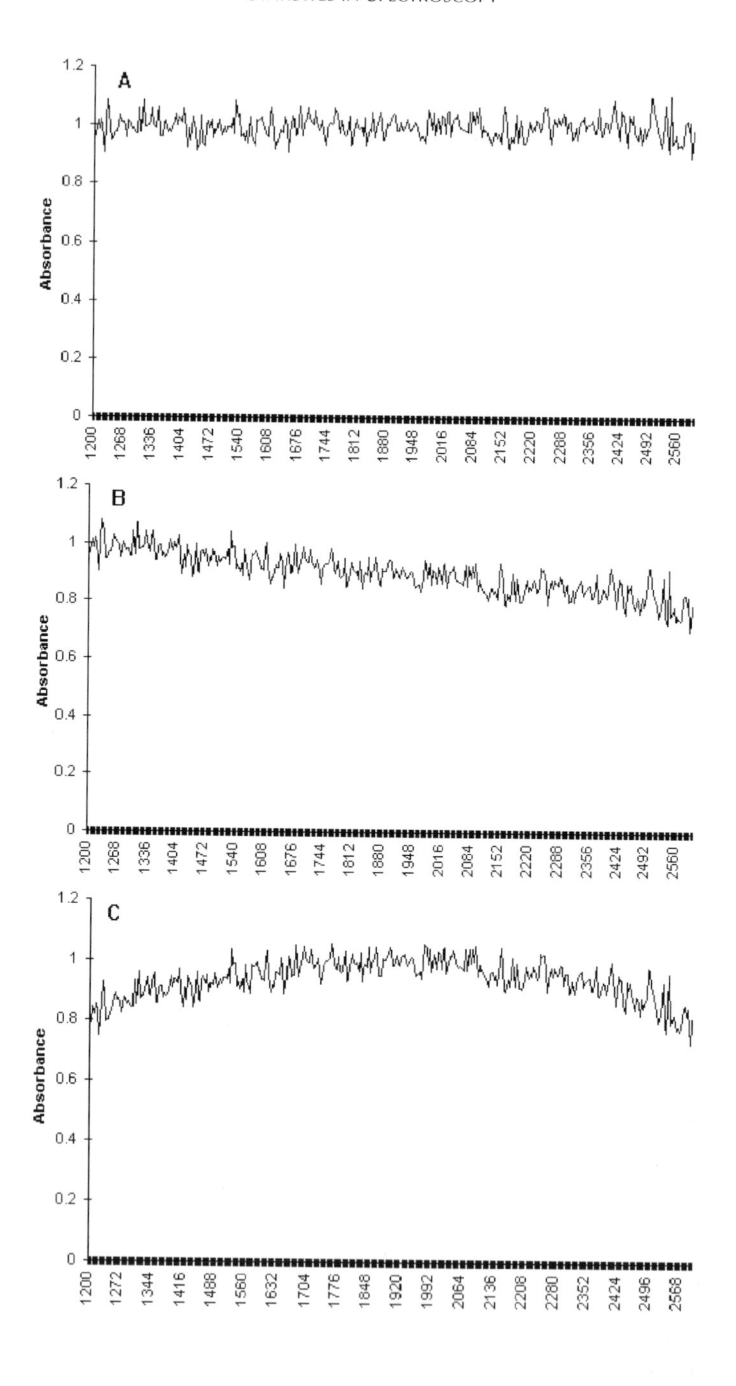
A
B
C
Absorbance
Absorbance
Absorbance
1.2
1
0.8
0.6
0.4
0.2
0
1200 1268 1336 1404 1472 1540 1608 1676 1744 1812 1880 1948 2016 2084 2152 2220 2288 2356 2424 2492 2560
1200 1272 1344 1416 1488 1560 1632 1704 1776 1848 1920 1992 2064 2136 2208 2280 2352 2424 2496 2568

flatness, but will estimate the noise level regardless of any slow wiggles it may exhibit. The method is based on the twin facts that, as we have shown, the difference between two data points estimates according to Equation 7-5, and these estimates can be pooled according to Equation 7-7.

If we were to take points at random from a 100% line that, in fact, was not flat, then the difference between those two points would also have a contribution from the tilt of the line. However, if we were to compute the difference between two adjacent points:

$$D = X_{i+1} - X_i$$

the tilt of the line would make little or no contribution to that difference; it would be a good estimator of the noise alone (or at least much better than computing it via Equation 7-1). If we then took several sets of adjacent pairs, and pooled the results, we would have an estimator of the noise that was immune to any variation of the data that occurred relatively slowly compared to the distance between the data points. How many such differences can we take? It turns out that we can use all $n - 1$ differences. Intuitively one might think that using a given point twice (once with the point before it, and once with the point after it) might invalidate the results. However, if the data are in fact random, then the differences between them are also random, and the difference between X_{i+1} and X_i is independent of the difference between X_i and X_{i-1}. Thus, the computation of standard deviation by the process of *successive differences* is a valid way of computing this statistic. The formula for this computation is:

$$S = \sqrt{\frac{\sum_{i=1}^{n-1}(X_{i+1} - X_i)^2}{2(n-1)}} \tag{7-8}$$

In the absence of systematic effects, this computation is an estimator of σ with $n - 1$ degrees of freedom just as Equation 7-1 is. Let us try these ideas out. Figure 7.1 shows a section of the 100% line used for illustration in Chapter 6, at 4 wavenumber intervals between 1200 and 2600 wavenumbers. Part A of Figure 7.1 shows the unmodified data. In part B we have added a linear tilt to the 100% line, and in Part C of Figure 7.1 we have added a parabolic curvature. Table 7-1 presents the data, while Table 7-2 shows the results of computing the standard deviation for the three cases, using the three methods of computation

FIGURE 7.1. Spectra corresponding to the data in Table 7-1. Figure 7.1-A is the 100% line corresponding to the data. Figure 7.1-B, representing a non-flat 100% line was generated by subtracting ($0.2 \times I/351$) from each data point, where I is the index number of the data point. Figure 7.1-C, representing a non-flat 100% line that cannot be corrected with a linear adjustment, was generated by subtracting ($(0.002 \times (I - 175.5)^2)/351$) from each data point. (A) Flat 100% line; (B) tilted 100% line; (C) curved 100% line.

TABLE 7-1

Data corresponding to the spectra of Figure 7.1. The data listed is for the flat 100% line in Figure 7.1-A. The data for Figure 7.1-B was generated by subtracting (0.2 × I/351) from each data point, where I is the index number of the data point. The data for Figure 7.1-C was generated by subtracting ((0.002 × (I − 175.5)^2)/351) from each data point.

0.95613	0.96794	1.01406	0.98738	1.01776	0.99777	0.90754	1.01685
1.08429	1.04230	0.94944	0.95113	0.97377	0.97562	1.00578	1.03619
1.01149	1.01043	1.00271	0.96033	1.01368	1.00482	0.98480	0.99249
0.97088	0.97159	1.05820	0.97478	1.08623	0.99382	0.99806	0.99961
1.01955	1.05854	1.00121	0.99136	1.03743	1.06349	0.96164	0.96265
1.01173	1.01584	0.97853	0.98753	0.97612	0.99567	1.01729	1.04011
1.00162	1.03022	1.02151	1.01234	1.05815	0.98643	0.92351	0.99098
0.94642	1.02727	0.99645	0.99055	0.91763	0.93781	1.03251	0.94026
0.93178	1.00572	1.01377	0.99203	1.01899	0.96024	0.98790	0.97995
1.01900	0.98992	0.96539	0.99561	0.97894	0.99613	0.99655	0.99230
1.02280	0.97243	1.08742	1.03108	1.03803	0.96076	0.97303	0.94973
0.98015	0.93725	1.02850	0.97338	0.95156	0.92857	1.02048	1.01558
1.01776	1.03104	1.00000	0.99784	0.97133	0.96236	1.03677	1.06301
0.97854	0.92113	0.94436	0.94262	0.98684	0.97947	0.97787	1.03044
1.00112	0.91094	0.97338	1.03560	1.02601	0.96562	0.97087	1.06775
1.01573	0.96526	1.00939	1.02796	1.06069	1.02636	1.00883	1.00767
1.05177	1.01631	0.98863	1.00170	1.02054	0.98250	0.93546	0.99011
1.00434	1.00960	1.00755	1.01475	1.06346	1.04127	0.99862	0.97744
1.02073	0.97209	0.96813	0.97672	1.03531	0.93499	0.94542	0.98773
1.02499	0.99775	0.99068	0.99703	0.96720	1.03029	0.93536	0.97724
0.99273	0.98367	1.04970	0.96093	0.99707	1.00809	1.05010	1.01677
0.95357	0.95121	1.01195	1.00447	1.00645	1.02759	1.04541	1.04001
0.98558	0.98775	1.01681	0.99678	0.98592	1.01237	1.02320	1.00226
0.98825	0.99903	1.01526	1.01076	0.97738	0.95503	0.96533	0.96969
0.94307	0.99546	1.05836	1.04860	0.99333	1.04718	0.97065	1.02777
1.01675	0.97815	1.03796	1.02729	0.97905	1.04869	1.05123	0.96573
1.00318	1.02642	1.02712	1.04444	1.01233	0.99565	0.99109	0.99333
0.98587	0.98174	1.05513	0.98053	1.05399	1.00984	1.05475	1.01268
1.06541	0.96876	1.00750	0.98689	0.98108	0.96010	0.95618	0.93728
0.98307	0.98416	0.96158	0.97568	0.94344	1.02562	1.07073	1.00780
0.92730	0.92542	0.96698	0.94697	0.98902	1.04335	0.94635	1.02341
0.99650	0.94414	0.94280	0.96156	0.97688	1.02029	0.98560	0.99938
0.99214	1.02392	1.01433	0.98484	1.00291	1.07006	1.06142	1.06607
1.02147	0.94487	0.99647	1.03092	1.00457	1.03205	1.02206	1.01244
1.04674	1.04013	0.98252	1.01710	1.01977	0.95727	0.96290	1.00236
0.95955	1.00000	1.01944	1.02662	1.03911	0.98946	1.00342	1.00820
1.01663	1.02225	0.98115	0.98292	1.01775	1.06812	0.97064	0.96998
0.98928	1.01946	1.01507	0.98518	1.00941	1.06679	1.09354	1.03385
1.01479	0.95602	1.05768	1.04374	1.04316	0.96644	0.93649	1.04325
1.01689	1.04144	0.98590	0.97228	0.95806	1.00333	0.98908	0.94379
1.00101	0.96911	0.98862	1.01234	1.10644	1.10574	1.06128	1.02270
0.99155	0.97962	0.93648	0.96976	1.02186	1.07674	0.94231	0.91421
1.10894	0.94546	0.95967	0.97866	0.93243	0.94422	0.93252	0.94659
1.01503	1.02222	0.98970	1.01795	0.89836	0.97717	1.01092	

TABLE 7-2
Standard deviation of 100% lines from Figure 7-1 computed three different ways.

	S.D. by Equation 7-1	RMS difference from fitted line	S.D. by successive differences
Flat 100% line (Figure 7.1-A)	0.0377632	0.0377920	0.0329188
Tilted 100% line (Figure 7.1-B)	0.0678936	0.0377921	0.0329197
Curved 100% line (Figure 7.1-C)	0.0656151	0.0657041	0.0329316

(A, the "normal" way, from Equation 7-1; B, by fitting a straight line (using $n - 2$ as the divisor), C, from successive differences.)

The results in Table 7-2 confirm our statements above: computation of standard deviation by the "normal" method, i.e., by using Equation 7-1, is indeed sensitive to both a linear tilt, and to curvature of the 100% line; for the departures from flatness shown in Figure 7.1, the standard deviation has increased by a factor of two for both cases of systematic effects.

Fitting a straight line to the data and computing RMS difference from the line does, as we expect, compensate for the linear tilt, but this computation does nothing to correct for the curvature. Indeed, since the curvature was added synthetically, the net slope over the 1200–2600 wavenumber region in Figure 7.1-C is exactly zero, and the computation of the differences from the fitted straight line is the same as the computation of standard deviation, except that we have thrown away one degree of freedom.

In the third column of Table 7-2 we have the result for computation by successive differences: we see that there is essentially no change of the computed standard deviation regardless of the lack of flatness.

The difference between 0.03292 and 0.03776, the values obtained by successive differences and via Equation 7-1 is a small difference. Nevertheless, in a book about statistics the reader would perhaps expect that we would test this question, especially since we have been spending so much time and effort discussing the concept of hypothesis testing. However, in this case, we have not yet addressed the question of an appropriate method for performing such a hypothesis test, so we leave that subject for a later chapter.

On the other hand, the difference between 0.032918 and 0.032932, the change in S.D. computed by successive differences on the flat and curved 100% lines, while so small as to be trivial even if shown to be not due solely to the effect of sampling random data, is of considerably more interest at this point. In this case hypothesis testing to determine if the difference is "real" is not appropriate. The reason in this case is that hypothesis tests determine the effect of random variations of the data, while in fact our statement is exactly true: the variations introduced into these data are systematic; there was absolutely no change made to the random part of the data. We will eventually learn how to use hypothesis testing to distinguish random from systematic effects, but we are not there yet.

Since the difference between the computation by successive differences and by the standard method is due to systematic effects, we know that the increase in standard deviation in this case is real, however small. In fact, some of the variation is due to the systematic effect "creeping" into the computation. We will eventually learn how to use the statistical technique of analysis of variance to separate the effects, and get an even better estimate of the noise level.

The careful reader will have noted above that we said that computation by successive differences results in an unbiased estimator of σ^2 only in the absence of systematic effects. When systematic effects are present, the result is a biased estimator, as the change in computed S.D. from 0.032918 to 0.032932 indicates. However, computation via Equation 7-1 also results in a biased estimator of σ^2 under these conditions, and comparison of the results in Table 7-1 should be convincing that the computation using successive differences is a much better estimator of σ in such cases than the standard computation.

Careful inspection of the first row of Table 7-2 also reveals that, in the absence of tilt in the 100% line, correcting for this systematic error when it does not exist results in a calculated value for that is too large, as the increase from 0.03776 to 0.03779 indicates. Again, even though this increase is small, it is real, and is due to the extra loss of a degree of freedom.

Now we have what seems to be a fundamental problem: if there is no systematic error, correcting for it results in the wrong value for sigma, as in the first row of Table 7-2; on the other hand if there is a systematic error, not correcting for it also results in the wrong value for sigma, as the second row shows. How can we handle this apparent dilemma?

Anyone who has read the last few chapters will have noted that no matter what subject we start to discuss, we always end up talking about hypothesis testing (notice how we snuck it in here as well). The reason for our obsession for that subject should start to become apparent now: hypothesis testing will eventually come to our rescue in cases like this: by telling us whether systematic errors exist, we will arrive at a means of deciding which method of calculation is the correct one to use.

Alternative Method #3

Finally, on the subject of 100% lines, we note that from inspecting the 100% line that we worked with in Chapter 6, we have reason to be suspicious that σ below 1000 wavenumbers is not the same as above 1000 wavenumbers. Examining that 100% line further also raises suspicions about the equality of σ above and below 3000 wavenumbers. To properly characterize the noise of the instrument, we should compute separately the noise in these spectral regions.

But this is not proper procedure. The division of the spectrum into these three regions is arbitrary, based upon subjective suspicions concerning the nature of the noise. To claim status as a *science*, we need an objective method of separating frequency-dependent phenomena. We have neither *a priori* reason to pick the number of regions nor the means of distinguishing them; consequently, the only objective approach is to treat each and every wavelength separately. But is there a means of estimating for each individual wavelength?

Of course there is, or we would not have brought the subject up. Clearly, what we need is an estimate of σ at each wavelength that contains many degrees of freedom. To obtain this, we should measure the 100% line many times. If we measure the 100% line m times, we can then apply Equation 7-1 to the data at each wavelength separately, to obtain a standard deviation with $m - 1$ degrees of freedom for each wavelength. Of course, this is a lot more work than measuring once and pooling data at the different wavelengths, but there is no reason an instrument's computer could not be programmed to do this automatically, and provide a "noise spectrum" for the instrument. Besides, who ever said good science was easy?

This approach will also be immune to systematic departures of the 100% line from flatness, as long as the systematic effects are reproducible. If these vary between measurements, then there is good reason to be suspicious that what is needed is a service call, not another way of measuring standard deviation.

Notes and References

7-1. "Specimen" is used rather than "sample" to designate a quantity of a material to be analyzed in order to avoid confusion with the statistical meaning of the word "sample". This nomenclature has been recommended by IUPAC, and working group 13.03.03 of ASTM has recently voted to recommend this nomenclature as an ASTM standard.

7-2. Youden, W.J., and Steiner, E.H., *Statistical Manual of the AOAC*, 1st ed., Association of Official Analytical Chemists, Washington, DC, 1975.

8

THE CENTRAL LIMIT THEOREM

Have we been lying to you all this time? No, not really: that was just a zinger to make sure you're paying attention. On the other hand, there are a few points where questionable doings are going on. On the one hand are sins of omission. These are O.K.; we simply have not yet gotten around to discussing some of the background needed to fill in the holes, and we will eventually get around to these.

On the other hand are sins of commission, or, at least apparent sins of commission. A careful reader might have noticed what appeared to be discrepancies between various statements we have made in different chapters. We wish to discuss one such apparent discrepancy, in order to clarify what is going on. The situation concerns the question of randomness.

In Chapter 2, we talked about the fact that, in order that the probabilistic concepts upon which the science of statistics is based remain valid, there is a requirement for randomness. We then defined a random process as one in which all items have an equal chance of being selected during any trial.

In other chapters, on the other hand, we have blithely discussed the Normal distribution, and other distributions where some values are more likely to be selected than others. Is it meaningful to talk about a random selection from these other distributions?

Well, let us go back and look at what we did. This all started by our taking a small set of uniformly distributed numbers, the integers 1–10, and taking a random selection of them. Clearly, in that case the selection was made so that every time we picked a number, each available number had an equal chance of being picked, so this conformed to our definition of randomness.

The distributions that gave rise to situations where some numbers were chosen more often than others were generated by adding together several of the (randomly) chosen numbers from the original set. Then we found that a few of the sums were selected more often than others (actually we used averages, but that is just a scaling factor: it is the summation that is important). The reason some sums were selected more often than others was because more combinations of original numbers gave rise to those sums more often than other sums. For example, if two of the original integers are being chosen each time and added, then there is only one way to get a sum of 20 (10 + 10), there are three ways to get 18 (10 + 8, 9 + 9, 8 + 10), six ways to get 15 (10 + 5, 9 + 6, 8 + 7, 7 + 8, 6 + 9, 5 + 10), and so forth. Thus, a sum of 15 (average of 7.5) is six times as likely to be obtained as a sum of 20, and three times as likely as a sum of 18. Thus, there is no conflict between our definition of randomness, which applies to the selection of numbers from the underlying set, and the non-equal appearance of results, which applies to the combination of randomly chosen numbers. A "random" selection from the non-uniform distribution really stands for the selection of one result from a set obtained by combining the underlying distribution, which selection is, in fact, the random one by our definition.

The important point from all this is that it is the combinatorial properties of the members of the initial set that determine the distribution of the results (the sum, in the case under discussion). The numbers themselves, and even their distribution, are less important than the combinations that they can form.

The basic assumption of statistical tests (e.g. Student's t, chi-square, Normal, F-test, etc.) is that populations composed of random events will follow some identified (most often Normal) distribution with known properties of mean and standard deviation. We have not yet proved that sets of measurements for a single analytical characteristic are always normally distributed, but this assumption generally approximates the truth. We note here that departures from normality for populations of measurements, events, or objects is not necessarily critical within the context of our specified uses of statistical tests. The reason is the same as above: this distribution arises naturally solely from the combinatorial properties of numbers, with very little regard to the numbers themselves.

We often assume that a Normal distribution applies when we are taking repetitive measurements on the same sample and when we are measuring the same characteristic for different samples. We must note here that when measuring the same constituent concentration in a variety of different samples, distributions other than Normal may be encountered. A frequently exhibited distribution in this case is the log-normal distribution. In the log-normal curve, the logarithm of the concentration of a set of samples gives a Normal curve when plotted against frequency (see, for example, Figure 5-2).

The appearance of various distributions comes from the way the underlying phenomena contribute, thus giving rise to the observation that some results are found more often than others, and always in certain well-established ratios, just as the averages of the integers occurred in completely defined ratios. In the case of adding integers, we found that the average (i.e., the sum scaled by a factor) tended to cluster around the population mean (μ), and that the clustering was tighter as more terms were included in the sum. The nature of the clustering, i.e., the function that determines the relative frequency of occurrence of the various possible sums, was undefined. We merely noted that more terms occurred closer to the population mean than far away, and the closer a number was to μ the more likely it was to occur.

Determining the functions that define the frequency of occurrence of various results occupied the attention of mathematicians for many years. Determining the distribution of the result of adding together numbers that are subject to error was the single most fundamental derivation in statistics, and the result is known as the central limit theorem (say, that even *sounds* important!)

History

We do not usually discuss the history of statistics here, but the central limit theorem is important enough to warrant some discussion of its background. The Normal distribution is often called the Gaussian distribution, implying its discovery by the mathematician Carl Friedrich Gauss. However, as is often the case with scientific discoveries, Gauss was not the first mathematician to discover this distribution. That honor belongs to the French mathematician Abraham DeMoivre who showed that, if many terms are added together, and the variation of each term (i.e., the error) has been randomly selected from an originally binomial distribution, then the distribution of the sum approximates the Normal distribution more and more closely as more and more terms are included in the sum. Other mathematicians then showed that the same held true if the original distribution was uniform, triangular, or any of several others. Eventually these specific cases were generalized, so that it was shown that the sum of terms from any particular distribution became approximately Normal if enough terms were included.

Finally, the French mathematician Pierre Simon de Laplace was able to prove the generalization that we now call the central limit theorem: he showed that the error of each individual term could come from any distribution whatsoever (i.e., the terms in the sum could come from *different* distributions), and the error of the sum would become approximately Normal if enough terms were used. There are only two restrictions that must be met in order for the central limit theorem to hold:

(1) The variance corresponding to any single term (error source) must be finite. This is never a problem with real data, although some theoretical distributions exist whose variances are infinite.

(2) The errors of each term must be approximately equal, and there must be enough terms included so that no one of them dominates. If this second condition does not hold, then the distribution of the sum will be the same as the distribution of the dominant term.

So why is this distribution called the Gaussian distribution? While Gauss did contribute to the formulation of the central limit theorem, he did so no more than other mathematicians of his time. Gauss' big contribution was in the use of this theorem: Gauss was the mathematician who showed that only if the errors were Normally distributed, would the least squares method of fitting an equation to data give the maximum likelihood that the fitted equation was correct (we will discuss this later in this book).

The key fact regarding the central limit theorem is just that: as long as each underlying error term is small, it does not matter what the individual distributions are; they need not even be the same: the resulting distribution will approximate the Normal distribution closer and closer as more individual error terms enter.

The reason for assuming that a Normal distribution applies when the nature of the distribution is unknown is not arbitrary (although the assumption may be made arbitrarily in particular cases). Fundamentally, what we are doing is assuming that, in the absence of information about the nature of the errors, we believe that any error we have in our data is due to the contribution of many small errors from different and unknown sources. Again, the combinatorial properties of the numbers are the only operative factors in determining the final overall distribution. Then, by the central limit theorem, the net error will indeed be approximately Normally distributed.

Derivation

The proof of the central limit theorem requires higher mathematics, well beyond the bounds that we have promised to limit our discussions to. Generally, this proof is also not included in most texts on applied statistics, for the same reason. For those interested in pursuing this, The references at the end of the chapter (8-1, 8-2) contain proofs of the central limit theorem with increasing rigor and generality (and difficulty!). Reference (8-1), in particular, is a delightful little book containing information, not otherwise easily found, on how to deal with errors in data.

On the other hand, the derivation of the central limit theorem is important enough that we cannot simply ignore its derivation completely. This is so not only because of the importance of the central limit theorem itself, but because this derivation is the prototype for the derivations of all the important distributions in statistics. We hope to convince our readers of the truth of statements made previously, regarding the mathematical rigor of the foundations of statistical operations. Thus, we will compromise. We will not actually derive the central limit theorem rigorously, but we will present highlights of the derivation, skipping over the hard parts, so that we can all see how these things are done. For this purpose we will follow the derivation of Beers (8-1).

The derivation starts, as such derivations do, by stating the conditions we wish to explain, and the fundamental postulates to be used. In the case of the central limit theorem, we wish to know the distribution of the sum of many terms, each term being subject to an independent error. The desired distribution is the distribution of probabilities of finding any given value for the sum. We then state some very general conditions: we assume that the distribution we wish to find does in fact exist, and that it has some well-defined form, although that form is not yet known. If these assumptions are true, then we can express the distribution as some yet-unknown function $f(z)$, of the distribution of values of z. We then let W be the probability of finding any given value of z between z and $z + \Delta z$:

$$W = f(z)\Delta z \tag{8-1}$$

But $f(z)$ is due to the contribution of many underlying terms, each term having its own (unknown) probability distribution $f(z_i)$. By the laws of probability (see Chapter 2), the net probability of independent items is the product of the individual probabilities, thus:

$$W = f(z_1)f(z_2)f(z_3)\cdots f(z_k)(\Delta z)^k \tag{8-2}$$

where k is the number of underlying terms (also unknown).

Because of the useful properties of the logarithm, we take the log of both sides of the equation:

$$\ln(W) = \ln f(z_1) + \ln f(z_2) + \ln f(z_3) + \cdots + \ln f(z_k) + k \ln(\Delta z) \tag{8-3}$$

Now it is necessary to introduce two more probabilistic assumptions, both of which have been previously discussed in earlier chapters:

The first assumption is that we are certain to get *some* value for a measurement we make, therefore the integral of $f(z)$ over all possible values of z (i.e., from minus infinity to plus infinity) is unity (see Chapter 2).

The second assumption is that the most probable value of a variable is the average. This follows from our discussion concerning biased and unbiased estimators: the average (arithmetic mean) is an unbiased estimator (see Chapter 3).

So far, so good. Here is where we get to the higher math. These two assumptions that we just made are introduced into Equation 8-3 in suitable mathematical form, and by so doing convert equation 8-3 into a differential equation:

$$\frac{\mathrm{d}(\ln(W))}{\mathrm{d}\bar{X}} = \frac{\mathrm{d}f(z_1)}{f(z_1)}\frac{\mathrm{d}z_1}{\mathrm{d}\bar{X}} + \frac{\mathrm{d}f(z_2)}{f(z_2)}\frac{\mathrm{d}z_2}{\mathrm{d}\bar{X}} + \cdots = 0 \qquad \text{(8-4A)}$$

The solution to this differential equation is achieved by expanding it into an infinite series, from which it is shown to be equivalent to a much simpler differential equation, still in terms of the unknown function $f(z)$ (this also requires a fair amount of "crank-turning": see Beers' book (8-1) for the details):

$$\frac{\mathrm{d}f(z)}{f(z)} = az\,\mathrm{d}z \qquad \text{(8-4B)}$$

where a is a constant whose value is not yet known. Note that in this equation, the unknown function $f(z)$ is one of the variables of the equation (if it helps, then mentally think of it as Y). The solution to this differential equation is:

$$\ln(f(z)) = \frac{az^2}{2} + \ln(C) \qquad \text{(8-5)}$$

where C is an arbitrary constant of integration.

That Equation 8-5 is the solution of Equation 8-4 can be verified by differentiating equation 8-5; Equation 8-4 will be recovered (use the substitution of Y for $f(z)$, and note that, since C is a constant, $\ln(C)$ is also a constant, and disappears upon differentiation). From the properties of logarithms, Equation 8-5 can now be put into its final form:

$$f(z) = C\exp(az^2/2) \qquad \text{(8-6)}$$

Whew!

Thus, the initially unknown function, $f(z)$, is indeed found to have a well-defined form. This form is forced on it by the properties that it must have, as couched in the probabilistic conditions introduced during the course of the derivation. As it stands, Equation 8-6 still has two unknown constants (a and C), but the functional form is now defined.

The confidence one can use in predicting the range of any statistical parameter of a set of data depends upon the assumed distribution type (e.g. normal, log-normal, Poisson, etc.), the number of samples used to calculate the statistic, and the variance of the data.

Parameters may, of course, be calculated from original sets of measurements or objects; but individual parameters can also be calculated from a set of all possible parameters for a given population. If we choose to calculate all the possible mean values from a population of 10 integers, we are able to generate a set of sample means with a specific distribution termed the "sampling distribution of the mean". The mean of this population of means is the same as the mean of the original population.

A measure of spread of this population of sample means is termed the "standard error of the mean". The sampling distribution of the mean will begin to approximate the normal distribution as the number of samples increases, regardless of the distribution of the original population. This result is so important it has a special name: it is known as the "central limit theorem". This theorem is truly *central* to many statistical calculations because it is the justification for assuming a Normal distribution in the absence of other information.

Sampling theory and the theory of distributions allow us to study the relationships between an infinite or very large population of samples or measurements and a small subset drawn from the population. These theorems allow us to estimate the unknown properties of a population by extrapolating measurements taken on a small population.

The theorems allow us to determine whether our observed differences between two samples or two sample subsets are due to chance or are due to "real" and significant differences. In spectroscopy, we may wish to determine how to represent our large populations of samples we will be analyzing with a small learning set used for calibrating an instrument; we may wish to determine if two spectra are really different, or if two sample subsets are from the same population or represent real differences and belong to separate populations. Many of these questions involve the use of hypothesis or significance tests. Any conclusions drawn from such statistical evaluations of data are termed inferences.

In spectroscopy, discussions of distributions could involve populations of: equations, prediction errors, calibration and prediction sample subsets, spectra, sets of instruments, and so on. Our knowledge and attitudes regarding a particular spectroscopic method or data set are based on a certain limited experience with a limited sample taken from a large population containing all possible measurements or samples of interest. We may, because of human nature, have a tendency to make definitive judgments on a set of data, a process, a technology, or a mathematical treatment based on a tiny subsample of all the possible data available.

We cannot predict the exact error involved in estimating population statistics from subsamples of infinite populations, as the true value of any population would have to be known to accomplish this. The precision of any measurement can be evaluated, though, if subsets are continually selected from a population and parameters estimated from these subsets. In order for us to draw any meaningful conclusions from the parameters calculated from a small subset of a large population, we must make assumptions regarding the distribution of the large population.

An interesting sidelight to this is that originally government censuses were taken to include *all* individuals within the population; small subsets of a population were suspect due to the lack of knowledge of distributions and the properties of random sampling. Hence, it was deemed necessary to measure the entire population (in both the statistical and demographic senses).

We know the basic principles and mathematical properties of selecting sample subsets from large populations. The known properties of sampling include:

(1) we are able to define (and enumerate) a distinct sample set;
(2) each sample of a set has a known probability (P) that it will be selected;
(3) we use a process of selecting individual samples so that the probability of selecting any individual sample is known;
(4) we can compute from a small subset of samples drawn from any larger set an estimate of some parameter found in the population (for example, $\bar{X}$ as an estimator of μ).

It is difficult to estimate any real properties of the large population if:

(1) the subsampling procedure is restricted (e.g., only aliquots from the top 1 cm of a sample jar are used);
(2) if a non-random or haphazard sampling technique is used (e.g., selecting those samples closest to the instrument);
(3) if samples are pre-judged as a method of selection (e.g., "Gee, this sample looks about right!");
(4) a real chemical difference was used to select samples (e.g., "these selected samples were easier to extract").

Thus, if any element of randomness is removed from a sampling technique, it precludes the use of random selection and/or potentially withdraws the validity of estimating population characteristics based on analysis of the population subsample. If we have a large enough subset from a large enough population, a parameter estimated from the subset will be normally distributed. As we have discussed previously in Chapter 5, the sample mean $\bar{X}$ estimates the true μ and the sample standard deviation (S.D.) estimates the true σ.

We can designate a notation for the expected value of any population statistic from a given subsample parameter estimator by:

$$E(S) = \sum_{i=1}^{n} \pi_i S_i = S \tag{8-7}$$

where $E(S)$ is the expected value of the parameter S, S is any population parameter (for example, σ, μ, etc.) π_i is the probability of obtaining the given value of S_i.

For example, if we calculate a sample mean ($\bar{X}$) and its corresponding sample standard deviation (s) from a population; how accurate is our estimate of μ and

σ? Well, from the properties of the Normal or Gaussian distribution, we can determine that approximately 1 in 3 sample mean values computed from a normal curve will differ from the true mean by more than 1σ. Likewise, we can calculate that only 1 in 20 sample means $\bar{X}$ calculated from a larger population of samples will exceed our true mean μ by more than 2σ, etc. We also know that we are able to compute *confidence intervals* using standard deviations.

Recalling our recognition of the marriage of statistics and probability, we compare our real experience with a limited subset selected from a larger population to the results we would expect given our assumed distribution model. Probability theory allows us to calculate our expected results, while statistics enable us to determine whether our real data is in agreement with what we had expected.

Thus, the greater the confidence with which we wish to assure ourselves that we have indeed included the population parameter within the confidence interval, the wider the confidence interval, and the less precisely we can pinpoint its true value. For example, with Normally distributed data (or errors), if we use a 95% confidence interval, we know that we include those results between -1.96 and $+1.96$ standard deviations; but if we want a 99% confidence interval, we only know that they lie within -2.55 and $+2.55$. It is analogous to the uncertainty principle: the more sure you are of including the result, the less you know about where it is. This works both ways: if we know the parameter (μ in this case) 95% of the measured data values will be within ± 1.96 standard deviations; conversely, if we do not know μ, we will find it within 1.96 standard deviations of our measured value that same 95% of the time.

Beginning

As a scientific tool, the proper use of statistics will allow us to determine whether our real data conforms to the expected distribution model, and thus informs us as to the correctness of our model and/or our data. In a critical sense, statistics gives us a measure of disagreement between observed and expected results. It allows us to determine the significance and sources of errors, deviations, and variations from our expected distribution model. Specific statistical tests are used for determining whether real data is in disagreement with expected results is by performing a statistical test. These tests (e.g., Student's t, chi-square, Normal, F-dist, etc.) will be addressed in later chapters.

References

8-1. Beers, Y., *Theory of Error*, Addison-Wesley, Reading, MA, 1962.

8-2. Cramer, H., *Mathematical Methods of Statistics*, Princeton University Press, Princeton, NJ, 1966.

9

SYNTHESIS OF VARIANCE

The sum of the squares equals the squares of the sums: true of false? Well, let us see. Consider the numbers 9 and 11. These can be written as:

$$9 = 10 + (-1) \tag{9-1}$$

$$11 = 10 + 1 \tag{9-2}$$

So, the squares of the sums can be written as:

$$(10 + (-1))^2 + (10 + 1)^2 = 9^2 + 11^2 = 81 + 121 = 202 \tag{9-3}$$

And the sum of the squares is equal to:

$$10^2 + (-1)^2 + 10^2 + 1^2 = 100 + 1 + 100 + 1 = 202 \tag{9-4}$$

Well obviously that was just a trick, wasn't it? Clearly we chose just those numbers that make it all work out, right? It cannot work for just any old numbers, can it? Or can't it? Let us take any old numbers and see. How about the first bunch of digits from pi? Let us break each number up as a sum with 39.5:

$$31 = 39.5 + (-8.5) \tag{9-5A}$$

$$41 = 39.5 + 1.5 \tag{9-5B}$$

$$59 = 39.5 + 19.5 \tag{9-5C}$$

$$27 = 39.5 + (-12.5) \tag{9-5D}$$

As above, the squares of the sums are:

$$31^2 + 41^2 + 59^2 + 27^2 = 961 + 1681 + 3481 + 729 = 6852 \qquad (9\text{-}6)$$

Now let us do it to the pieces to get the sum of the squares:

$$\begin{aligned} &39.5^2 + (-8.5)^2 + 39.5^2 + 1.5^2 + 39.5^2 + 19.5^2 + 39.5^2 + (-12.5)^2 \\ &\quad = 1560.25 + 72.25 + 1560.25 + 2.25 + 1560.25 + 380.25 \\ &\qquad + 1560.25 + 156.25 = 6852! \end{aligned} \qquad (9\text{-}7)$$

Remarkable!

Well, is it a trick or isn't it? Where did the 39.5 come from? Is that the trick? Is there a trick? O.K. what is the trick?

It is really all very simple (is not everything, once you know it?) The number 39.5 just happens (and not by coincidence, either) to be the mean of the numbers 31, 41, 59 and 27. This property of sums can be proven to be general, and we will now proceed to do so:

The Proof

Consider a sum of the squares of any arbitrary set of numbers X_i (we leave out the limits of summation in order to simplify the expressions):

$$\sum X_i^2$$

Let us add and subtract the mean of the numbers from each number; the sum then becomes:

$$\sum X_i^2 = \sum (X_i + \bar{X} - \bar{X})^2 \qquad (9\text{-}8)$$

Expanding this, without canceling any terms, we get:

$$\sum X_i^2 = \sum (X_i^2 + 2X_i\bar{X} - 2X_i\bar{X} - 2\bar{X}^2 + 2\bar{X}^2) \qquad (9\text{-}9)$$

Whew, what a mess!

Canceling terms selectively, to get rid of the constants (i.e., the 2's), we then rearrange the terms:

$$\sum X_i^2 = \sum ((X_i\bar{X} - \bar{X}^2) + X_i^2 - X_i\bar{X} + \bar{X}^2) \qquad (9\text{-}10)$$

After distributing the summation, the expression in the inner parentheses can be factored and rewritten as: $\sum \bar{X}(X_i - \bar{X})$. Since $\bar{X}$ is a constant, it can be taken out of the summation, leading to:

$$\sum (X_i\bar{X} - \bar{X}^2) = \bar{X} \sum (X_i - \bar{X}) \qquad (9\text{-}11)$$

But, by the nature of $\bar{X}$, $\sum(X_i - \bar{X}) = 0$, therefore, the term in the inner parentheses of Equation 9-10 disappears, leaving:

$$\sum X_i^2 = \sum(X_i^2 - X_i\bar{X} + \bar{X}^2) \qquad (9\text{-}12)$$

It now becomes convenient to distribute the summations and rearrange the terms:

$$\sum X_i^2 = \sum \bar{X}^2 + \sum X_i^2 - \sum X_i\bar{X} \qquad (9\text{-}13)$$

Factoring the constant ($\bar{X}$) out of the last term, we convert Equation 9-13 to:

$$\sum X_i^2 = \sum \bar{X}^2 + \sum X_i^2 - \bar{X}\sum X_i \qquad (9\text{-}14)$$

Noting that $\bar{X} = \sum X_i/n$, or conversely, $\sum X_i = n\bar{X}$, Equation 9-14 becomes:

$$\sum X_i^2 = \sum \bar{X}^2 + \sum X_i^2 - \bar{X}(n\bar{X}) \qquad (9\text{-}15)$$

$$\sum X_i^2 = \sum \bar{X}^2 + \sum X_i^2 - \bar{X}(2n\bar{X} - n\bar{X}) \qquad (9\text{-}16)$$

Using the fact that $\sum \bar{X} = \bar{X} + \bar{X} + \bar{X} + \cdots = \bar{X}(1 + 1 + 1 + \cdots) = \bar{X}\sum(1) = n\bar{X}$, in conjunction with the definition of $\bar{X}$ yields:

$$\sum X_i^2 = \sum \bar{X}^2 \sum X_i^2 - \bar{X}\left(\sum 2X_i - \sum \bar{X}\right) \qquad (9\text{-}17)$$

$$\sum X_i^2 = \sum \bar{X}^2 + \sum X_i^2 - \sum \bar{X}(2X_i - \bar{X}) \qquad (9\text{-}18)$$

$$\sum X_i^2 = \sum \bar{X}^2 + \sum(X_i^2 - 2X_i\bar{X} + \bar{X}^2) \qquad (9\text{-}19)$$

and finally:

$$\sum X_i^2 = \sum \bar{X}^2 + \sum(X_i - \bar{X})^2 \qquad (9\text{-}20)$$

Q.E.D.

So, using only algebra (a lot of it, to be sure) we have shown that, under the proper conditions, the sum of the squares does indeed equal the squares of the sums—and our fingers never left our hands. The key to this separation of sums of squares is the step from Equation 9-11 to Equation 9-12. While no single difference from the mean is zero, the fact that $\sum(X_i - \bar{X}) = 0$ disappears (note—yes, we know it is intransitive) the cross product, and allows the equality to exist (exercise for the reader: prove this, i.e., that $\sum(X_i - \bar{X}) = 0$.

The Why

This process, which we have derived for the simplest possible case, is known as *partitioning the sums of squares*, and is the underlying mathematical foundation for the statistical procedure known as analysis of variance. Why? Look at equation 9-20. If we divide both sides by $n - 1$:

$$\frac{\sum X_i^2}{n-1} = \frac{\sum \bar{X}^2}{n-1} + \frac{\sum (X_i - \bar{X})^2}{n-1} \tag{9-21}$$

we see that the square root of the second term on the right-hand side of Equation 9-21 is the standard deviation of the X_i, and the second term itself, the square of the standard deviation, is known as the variance.

The derivation presented is the simplest possible case of the partitioning of the sums of squares, and in itself is of little interest except for its pedagogical value. The pedagogical value comes when we note that the two terms on the right-hand side of Equation 9-20 are the sums of squares associated with the mean and the standard deviation, respectively. The really important point about the partitioning of the sums of squares, and the practical utility of this procedure comes from the fact that if the necessary information is available then the term $\sum (X_i - \bar{X})^2$ can itself be further partitioned into other sums of squares. The methods devised to collect the necessary information have been collected under the concept known as "statistical design of experiments", a subject which we will not consider until later.

The fact that the sums of squares can be further partitioned is of supreme importance not only to statisticians (who hardly ever generate their own data, anyway), but also to spectroscopists and chemists in general, because it is this procedure that allows an objective evaluation of an analytical method. The reason for this is that when the calculation of sums of squares is performed on the data that represent the errors of an analytical method, it then becomes possible to allocate error contributions to different physical error sources. How? Look at this:

Using a shorthand notation, we let TSS represent the total sum of squares $= \sum (X_i - \bar{X})^2$, then we can write the partitioning of the sums of squares as:

$$\text{TSS} = \text{SS}_1 + \text{SS}_2 + \cdots \tag{9-22}$$

where the various SS_i each represent the sum of squares due to different error sources.

Dividing each term by degrees of freedom converts each sum of squares into a variance, just as we did in Equation 9-21. Thus, the total variance is equal to the sum of the variances from each error source. Using S^2 to represent the variance, we then see that:

$$S_\text{T}^2 = S_1^2 + S_2^2 + \cdots \tag{9-23}$$

This adding together of variances to determine the total is the synthesis of variance of the chapter title. The standard deviation corresponding to each source of error is then designated by S, thus leading to:

$$S_{\mathrm{T}} = \sqrt{S_i^2 + S_2^2 + \cdots} \tag{9-24}$$

The Wherefore

We will consider a hypothetical example, typical of our own experience in near-infrared reflectance analysis. As with all analytical techniques, this is subject to a number of sources of error, typical ones being electronic noise, reflectance differences upon repacking samples, and sampling error due to non-homogeneity of the material being analyzed. In addition, instruments are usually calibrated against a set of training samples that each have assigned values determined by a reference method (often a wet chemical procedure). We will consider how to determine the error associated with each error source in later chapters: it is a big topic); for now, let us assume that we have measured the error associated with each source and the standard deviations are as follows:

Reference method: 0.2%
Sampling: 0.1%
Repack: 0.05%
Electronic noise: 0.015%

We can then estimate the total error of the analysis as:

$$S_{\mathrm{T}}^2 = 0.2^2 + 0.1^2 + 0.05^2 + 0.015^2 \tag{9-25}$$

$$S_{\mathrm{T}}^2 = 0.04 + 0.01 + 0.0025 + 0.000225 = 0.0527 \tag{9-26}$$

The standard deviation of the analysis, then, is

$$S_{\mathrm{T}} = \sqrt{0.527} = 0.2296. \tag{9-27}$$

There are two rather subtle points that need to be made before we can consider the meaning of Equation 9-25. Again, we bring them up because ignoring the statistical requirements of calculations such as these is the pitfall that we see people falling into time and again, but space limitations preclude more than a brief mention; we must defer deeper consideration until a future chapter.

The first point that needs to be made is that the errors due to the different sources must be uncorrelated. While the partitioning of sums of squares is an exact algebraic procedure as we have seen, and the variances will also add with algebraic exactitude, nevertheless if the errors are correlated, then the sum will be a biased estimator of the true total population variance, rather than an unbiased one, and so will the estimates of the errors due to the individual phenomena. Note that this does *not* mean that the measured correlation coefficients among the actual measured data must be zero. Rather, it means that the *population* value of the correlation coefficient (called ρ by statisticians) between the individual error sources must be zero. How can we tell what that is? Sometimes physical theory allows an estimate of the correlation among the readings to be made, otherwise usually it is unknown. In either case, however, proper procedure dictates that before proceeding with an analysis of variance, a hypothesis test on the correlations should be made. If the null hypothesis on the statistic for correlation ($R = 0$) is confirmed, then it is permissible to assume that the population value is confirmed ($\rho = 0$).

The second point that needs to be made is that we have spoken of the total sum of squares (or rather the error computed from the total variance) as the error of the analysis. Strictly speaking, it is no such thing; it is merely the total sum of squares. The error of the analysis must be determined by comparing the analytical values to the "true" values of the materials being analyzed (note: not necessarily even the population values, the population value of a given lot of material is not necessarily the same as the "true" value of the sample of that lot against which the calibration calculations are being performed). The "true" values are, however, unknown. We only have the values from the reference method to compare with. Thus, measures of error in cases such as this do not truly measure the error of an analytical method. Often, especially when multivariate (i.e., chemometric) methods of calibration are used, the error is computed as:

$$\sqrt{\frac{\sum_{i=1}^{n}(X_{\text{test}} - X_{\text{ref}})^2}{n - 1}}$$

and that is considered the error. But this expression really corresponds to S_{T} in Equation 9-24.

Now we are ready to consider Equation 9-25. We note that, while the errors of NIR analysis from each of the listed sources are more or less comparable (certainly the first three), the variance due to the reference laboratory error is 0.04/0.0527 = 75.9% of the total variance. This is common: the combining of variances usually causes the largest error source to be overwhelmingly the dominant error of the analysis. For this reason, the greatest improvement in analytical accuracy can be achieved by working on the largest error source alone (you already knew that, but now we can quantitate it).

For example, if sampling error could be completely eliminated, the best we would expect to do would be:

$$\sqrt{0.04 + 0 + 0.0025 + 0.000225} = 0.206 \qquad (9\text{-}28)$$

On the other hand, if the reference error were reduced by a factor of $\sqrt{2}$, to $0.2/\sqrt{2} = 0.1414$, then the total error would be:

$$\sqrt{0.02 + 0.01 + 0.0025 + 0.000225} = 0.181 \qquad (9\text{-}29)$$

Thus, reducing the largest source of error by half (in terms of variance) provided more benefit than eliminating the second-largest error entirely. Moral: Less effort is needed to achieve better results, when the effort is directed to the proper place (obviously there is an assumption here that all errors are equally easy to reduce).

When performing calculations such as this, it is important to watch the dimensions of the numbers. We have noted before (in Chapter 6) that, when doing hypothesis tests, and when comparing variances, it is invalid to compare members of different populations. Similarly, when synthesizing variances, it is also invalid to add variances of different populations. Often, however, a transformation of the data will allow comparable variances to be obtained. This is done via standard mathematical treatment of the data, assuming that the relations are known from physical theory. Known as "propagation of error", this is discussed in several books (see Reference (9-1), for example). The general method uses the fact that if $y = f(x)$, then $\mathrm{d}y = \partial f(x)/\partial x \, \mathrm{d}x$. We consider one simple case here, as an example.

In quantitative spectrophotometric work, the electronic noise shows up as an error on the transmittance or reflectance value. The reflectance we will call r, the error we will call S_r. However, the quantitation is usually done on absorbance data, where the absorbance is defined as $A = -\log_{10}(r)$.

Therefore,

$$S_\mathrm{A} = \frac{\partial A}{\partial r} S_\mathrm{r} = 0.434 \frac{\partial \ln(r)}{\partial r} S_\mathrm{r} = 0.434 \frac{S_\mathrm{r}}{r}$$

As a numerical example consider a measurement where $r = 0.810$ with a standard deviation, following several measurements, of 0.005. Then $A = -\log(0.81) = 0.092$. $S_\mathrm{A} = 0.005 * (-0.434/0.81)$ which results in $S_\mathrm{A} = 0.00268$.

Prelude to Analysis of Variance

Having shown how to synthesize variances, we now turn the tables around and give a preview of how and why statisticians have also learned to take them apart. While we defer further consideration to later chapters, here we note that that statistical operation, which is called "analysis of variance" (often abbreviated as ANOVA) is based, in its simplest form, on Equation 9-20, which we rewrite here with an addendum:

$$\sum_{i=1}^{n} X^2 = \sum_{i=1}^{n} \bar{X}^2 + \sum_{i=1}^{n} (X - \bar{X})^2 = n\bar{X}^2 + \sum_{i=1}^{n} (X - \bar{X})^2 \tag{9-30}$$

where $\bar{X}$ is the mean of the X_i, and the summation is taken over all the samples in the set. The proof of Equation 9-30 was presented above. The key to the proof of this equality is the fact that the cross-product term: $\sum(X - \bar{X})$ is always zero, because of the nature of $\bar{X}$. Since Equation 9-30 is generally true, it is also generally true that, when the squares are suitably formed, the sum of the squares equals the squares of the sums. This is one of the major reasons why statisticians prefer to use variances rather than standard deviations in computations: variances add. This leads to an interesting possibility: if variances add, there is the potential to determine the individual variances that contribute to the total. It is this procedure of finding the individual variances that constitutes the process of analysis of variance.

Why would we be interested in finding the individual variances? In at least one particular case the answer to this question is easy to give: if the total variance is the error of an analysis, and this total error is the result of the effect of several different error sources on the measurement. In this case, determining the magnitudes of the individual error sources will reveal which of the many phenomena involved is the major contributor. Then, the available resources can be used to study and presumably reduce the effect of the phenomenon contributing the most error. In this way resources are used most efficiently. Let us start by working backwards, and see what happens if we synthesize variance. For example, suppose that a measurement is effected by two sources of error, one of which, by itself, would cause the readings to have an S.D. of 0.25 (let us consider all standard deviations to be population values, σ) the other, by itself, would give rise to an S.D. of 0.50 (we do not list the units, because it does not really matter what they are, as long as they are the same for both cases). The variance of the result is $0.5^2 + 0.25^2 = 0.3125$. Then the total error of the result is $\sqrt{0.3125} = 0.559$. Now, if the two phenomena giving rise to these errors were equally amenable to study, then clearly we would wish to spend our time (and resources, and energy, and money(!)) on finding ways to reduce the second error source. After all, if we eliminated the first error entirely, we would reduce the total error to 0.5. On the other hand, if we could reduce the magnitude of the effect of the second error source by only half, the total error would be reduced to $\sqrt{0.25^2 + 0.25^2} = 0.353$. Thus, reducing the larger error by half (which presumably would require only half the effort) gives more benefit than eliminating the smaller error entirely. Thus, it is clearly beneficial for us to have a way of determining the contribution of the different error sources to the total error we see, and the tool for performing this operation is analysis of variance.

In our next chapter we will take a breather, review where we have been and maybe even look ahead a bit to where we are going.

References

9-1. Miller, J.C., and Miller, J.N., *Statistics for Analytical Chemists*, 1st ed., Wiley, New York, 1984.

10

Where Are We and Where Are We Going?

Well, we feel it is time to take a look around, to see where we have been, and where we are. This also gives us the opportunity to discuss some miscellaneous subjects that do not fit in very well anywhere else.

In the last nine chapters we covered very basic concepts of probability and statistics; discussion of these topics is somewhat unusual for chemists, who are not mathematicians. As we review the bulk of statistical literature there is a great deal written within the fields of biology, psychology, and economics, but little has been published dealing with the statistical aspects of data analysis in chemistry (with perhaps the historically notable exception of W. J. Youden (10-1)). Only a small amount of material is available specific to statistical methods and spectroscopic data.

The general complaint among classical spectroscopists against the use of complex mathematical interpretation of data has been previously expressed by Dr. (Hon.) Samuel Clemens when he related something like "There are three kinds of lies: there are lies, [bleep] lies, and statistics." The purely empirical approach with little or no theoretical explanation has been disturbing to real analytical chemists. Until more fundamental truths can be demonstrated using sound statistical methods and substantive chemical/spectroscopic theory, it has been suggested we would be better off not betting next month's rent on the numbers generated using purely empirical chemometric devices. Of course the picture is really not so bleak (maybe a little anemic, but not bleak), as there is usually some *a priori* chemical/statistical knowledge of our samples. On the other hand, random variation of data is a reality, and it is no solution to ignore the field of science that teaches us how to properly deal with such variations, simply because it has been misused. Also, the serious user is cautioned against letting a computer assume the role of the bus driver as it displays, "Just sit back and leave the thinking to me!" Therefore, this book is a basic "driver training" course so that we as chemists will be able to plot our own scientific route and begin to understand where we are going.

This chapter, then, is a short one; its main purpose is to summarize the previous ones and collect our thoughts regarding the next several chapters. The knowledge required to correctly apply the modern tools of chemometrics to real spectroscopic problems is specialized and requires a convergence of several disciplines including: chemistry, optics, mathematics, statistics, and computer science into a single unified subdiscipline. To date we have incorporated several of these disciplines while trying to maintain a discussion specific to statistics in spectroscopy. Topics to date have included: random numbers, probability theory, populations, degrees of freedom, distributions in general, the normal distribution, standard deviation, the central limit theorem, and variance. During the course of these chapters we introduced the key concepts underlying three statistical analyses of data. The single most important of these is the simple fact that, when random phenomena are involved, the behavior of the data can be described only in probabilistic terms.

Knowing the distribution of values produced by any given random phenomenon allows us to tell whether a measurement has been subjected to external influences; this concept underlies the procedure of hypothesis testing. Hypothesis testing is the one procedure that should be learned above all others: Before you try to explain something, make sure there is something to explain. Performing a hypothesis test, and finding that the results cannot have arisen solely by the chance random variations of the phenomenon affecting the data, verifies that there is, indeed, something to explain. Otherwise, there is the danger of trying to "explain" the randomness of the data.

In our introductory chapter we brought up practical questions such as: "When are two spectra the same?"; "How can I determine the contribution of any given error source to my final answer?"; and "What is the meaning of the published noise level of my instrument?" The science of statistics, by properly applying the mathematical descriptions of random behavior to measured data, allows objective answers to these questions to be formulated. We have started on the road leading toward the answers to those questions, although we have proceeded farther on along some paths than along others.

In the initial publication of this book as a series of journal columns we received the results of the reader response cards. The response to what we wrote so far was fairly good: out of about 80 reader response cards that we received, 77 were favorable, one thought we were being too simple, one thought we were too advanced and one did not think the subject warranted so much journal space. We seem to have hit a good average!

Chapters to come will advance us toward our goal of developing statistical theory to the point where we can answer the above questions, along with others. Some of these other questions, which we will also address might be, "How many times must an experiment be repeated in order to ensure reliable results?"; "Why do we use the least-squares criterion to calculate calibration lines?"; "What are the underlying advantages and disadvantages of multiple linear regression?"; "What are nonparametric chemometric techniques and what is their usefulness relative to parametric methods?"; What are the important aspects of the different mathematical calibration methods?"; "How do I test for significance using chemometric methods?"; and the like. Other topics are to deal with the Student's t-test, the use of statistical tables, Chi-square, the F-statistic, curve fitting, the value of correlation, specialized statistics, identification of outliers, discriminant analysis, and others.

Similar questions can be raised concerning the new multivariate methods of data analysis that are currently in vogue: What, for example, is the confidence interval for a principal component? What is the correct method to compare several different multivariate methods?

Some methods, such as simultaneous multiple linear regression (often abbreviated as MLR), are older techniques and have been highly developed by statisticians, so that answers to such questions are known. Others, such as principal component analysis and partial least squares are newer and have not had the benefit of similar extensive development, so that answers to such questions are, as yet, unknown. However, the questions remain; but except for the "voice in the wilderness" of Ed Malinowski (10-2), articles on neither the theory (10-3) nor practice (10-4) of chemometrics even acknowledge the existence of these limits to our knowledge.

While everybody speaks glibly about "noise" and "signal to noise ratio" in spectroscopic measurements, lack of appreciation of the effect of noise on measurements leads to some strange statements: in a recent article in *Applied Spectroscopy*, for example, variation in the computed position of a weak shoulder led the author to make the following statement:

"... the instrument appears stable, and the variation probably originates in the peak picking algorithm" (see page 1103 in (10-5))

It is hard to imagine an algorithm being unstable, or varying unless a different algorithm is used each time the computation is performed. Much more probable (and, we suspect, what the author really meant) is the likelihood that the small shifts in computed peak position are due to the sensitivity of the algorithm to the small variations, caused by noise, in the data used. A prior assessment of the sensitivity of the computed peak to the inherent instrument noise level, and a determination of the distribution of peak positions obtained when the instrument was known to be performing properly, would allow confidence limits to be placed on the allowable variation of the computed peak. We do not wish to be hard on Dr. Smith; earlier in the same article he does correctly indicate the use of statistical quality control procedures to keep track of an instrument's performance, in addition to the other fine work he has done. Rather, we want to emphasize the need to be continually on guard against attributing the effects of random noise to other phenomena.

Speaking of non-parametric versus parametric statistics, as we are about to do, allows us to introduce a discussion of our approach to this subject. We spent a good portion of one chapter discussing the meanings of the word "Statistics". One thing we did not note then was the fact that the science of statistics can be divided into two branches: "mathematical statistics" and "applied statistics". Mathematical statistics concerns itself with the derivations and proofs of the mathematical constructs used. For the most part, we avoid these, except where, in our opinion, they have pedagogical value either in helping understand how and why things are done in a particular way, or in convincing our readers that there is, indeed, a sound and rigorous basis to what is going on. Applied statistics concerns itself with the proper way to use the mathematical constructs in analyzing data. As such, it not only helps us to know what to look for in data, but also how to recognize it, and, perhaps, what to do about it. The key to all this is the concept of a parameter (hence "parametric" statistics), which we have already introduced. The underlying assumptions are that a population can be defined, and that for that population, a parameter corresponding to an item of interest does, in fact, exist. Statistics are then calculated in such a way that they are, to use formalistic terminology, maximum likelihood, unbiased, minimum variance estimators of the parameters.

The use of the term maximum likelihood reflects the origins of the science of statistics in probability theory, and is also descriptive of the nature of the statistic: witness that, in all the cases we have studied so far, the arithmetic mean is, in fact, the *most likely* value to find when a random sample is selected from the population of interest. Desirable statistics are unbiased, so that they do, in fact, estimate the corresponding parameter; parametric statistics are defined so as to be unbiased. Minimum variance: it can be shown that, of all the possible ways to estimate a parameter (e.g., the median and the mode of a sample, as well as the sample mean, are also unbiased estimators of the population mean) the ones defined and used as "parametric statistics" will have a smaller variance than any other possible estimator of the corresponding parameter, making the parametric statistics the best possible estimator.

Why, then, are non-parametric statistics used at all? The shortcoming of parametric statistics is that many of them are known, i.e., their distributions and confidence levels are known, only when the distributions of the underlying populations are of certain specified forms. Most commonly, the underlying populations must have a Normal distribution. While the existence of this condition is often assured by the central limit theorem, if a suspicion exists that the condition fails to obtain, then use of non-parametric statistics will give results that are robust against the failure of the underlying assumptions to be met.

A topic that more or less enhances our basic discussions of degrees of freedom, population sampling, and estimation of parameters can be incorporated into a single question: What is the effect of sample size on our precision and accuracy? There is a general trade-off between precision and reliability, which can be stated as an inverse relationship. As we have seen before, when we give up some precision we stand a much greater chance of being right. We are implicitly defining the term precision to be the width of the confidence interval for any given statistic, and reliability as related to our confidence level. We will consider this subject further in the next few chapters, and learn that, if we wish to increase both precision and reliability we must increase the size of our sample. We will learn the relationships between the precision (confidence interval), reliability (confidence level), determination of the correct critical values to use, and the number of samples needed to achieve the desired goals.

Then we may be able to agree with Polonius' assessment of statistics as he tells Hamlet, "Though this be madness, yet there is method in't."

References

10-1. Kruskal, W., and Tanur, J., *International Encyclopedia of Statistics*, The Free Press, Division of MacMillan Publishing Co., Inc., New York, 1978.

10-2. Malinowski, E., *Journal of Chemometrics*, 1, (1987), 33–40.

10-3. Beebe, K., and Kowalski, B.R., *Analytical Chemistry*, 59:17, (1987), 1007A–1017A.

10-4. Robert, P., Bertrand, D., Devaux, M.F., and Grappin, R., *Analytical Chemistry*, 59, (1987), 2187–2191.

10-5. Smith, A.L., *Applied Spectroscopy*, 41:7, (1987), 1101–1105.

11

MORE AND DIFFERENT STATISTICS

In this chapter we continue with the development of statistical theory. In previous chapters we have demonstrated the following facts:

(1) The average of samples from a population ($\bar{X}$) clusters around the population mean μ more and more tightly as more samples are included in the average (see Chapter 3).
(2) The technique of analysis of variance shows that when different sources of variability act on a datum, the total variance (square of the standard deviation) of the result is equal to the sum of the variances due to each individual source (see Chapter 9).

So we have shown that variances from different sources add. How about variances from the *same* source? For example, if we add up a set of numbers, each number containing a source of variance in the form of error, what is the variance of the sum?

The two facts listed above lead us to look for the answer to the question along the following train of thought: when more than one effect is operative, the variance of the result is the sum of the contributions from the variance of each effect. The variance of the result of only one effect is, of course, the variance of that effect. When the result is subject to two effects, the variance of the result is the sum of the variances of the two effects, and so forth.

Just as variances add when the sources are heterogeneous, so do they add when the sources of variance are homogeneous, as long as the source(s) of variability causing the changes of each number are all random and mutually independent. As the considerations of analysis of variance showed, the sum of the samples will have a variance equal to the sum of the variances of the samples. We can write this as the following formula (recall that the variance (V) is the square of the standard deviation, this allows us to simplify the expressions by leaving off the radical). Since the sources of variance are all the same, they have the same variance each time, therefore $V(X_i)$ is constant and:

$$V\left(\sum X_i\right) = \sum V(X_i) = nV(X_i) \tag{11-1}$$

where $V(X_i)$ is read as "the variance of X_i". Equation 11-1 follows from the fact that the variance of each X_i estimates the *same* population variance.

We now need to know the variance of a multiple of a random variable (X is the variable and a is the multiplier, a constant). We approach this by computing the variance of a number of values of aX by simply plugging aX into the equation for standard deviation:

$$V(aX) = \frac{\sum_{i=1}^{n}[(aX_i) - (\overline{aX})]^2}{n-1} \tag{11-2}$$

Since a is a constant, $a = \bar{a}$ (exercise for the reader: prove this), and Equation 11-2 becomes:

$$V(aX) = \frac{\sum_{i=1}^{n} a^2X^2 - 2a^2X\bar{X} + a^2\bar{X}^2}{n-1} \tag{11-3}$$

$$V(aX) = \frac{\sum_{i=1}^{n} a^2(X^2 - 2X\bar{X} + \bar{X}^2)}{n-1} \tag{11-4}$$

$$V(aX) = a^2\frac{\sum_{i=1}^{n}(X - \bar{X})^2}{n-1} \tag{11-5}$$

$$V(aX) = a^2\mathrm{Var}(X) \tag{11-6}$$

Now, the average is $1/n$ times the sum; therefore, replacing a by $1/n$ in Equation 11-6 gives the expression for the variance of the average:

$$V(\bar{X}) = \left(\frac{1}{n}\right)^2 V\left(\sum_{i=1}^{n} X_i\right) \tag{11-7}$$

Using the value for the variance of a sum from Equation 11-1:

$$V(\bar{X}) = \frac{1}{n^2}nV(X_i) \tag{11-8}$$

and finally:

$$V(\bar{X}) = \left(\frac{V(X_i)}{n}\right) \tag{11-9}$$

Taking square roots on both sides gives the more or less well-known result:

$$S.D.(\bar{X}) = \frac{S.D.(X)}{n^{1/2}} \tag{11-10}$$

Equation 11-10 gives a numerical value to what we have noted only qualitatively until now, and expressed in words in Point 1 above: not only do the averages of samples cluster around μ; we now have proven that the clustering tightens with the square root of the number of samples in each average.

In other words, if we go back to Table 3-2, and calculate the standard deviations of each set of 20 means (each mean itself being the average of n random samples from the population of integers 1–10), those standard deviations should decrease with the square root of the number of integers (taken from the population of the integers 1–10) used in each set of averages. The results obtained by actually doing that are presented in Table 11-1, and these pretty much are in accordance with the equations we have just derived, at least within limits we might reasonably expect, considering that these standard deviations are themselves statistics, and subject to variability. We will shortly learn how to determine formally whether our expectation is correct.

In addition to the two facts listed above, we have also learned another important fact: in addition to the variance of a sum increasing as n V(X),

TABLE 11-1
Standard deviations of averages of n samples from the population of integers 1–10. The list of averages appears in Table 3-2.

n	S.D. of 20 means each mean calculated from n samples
1	2.903718125
2	1.580098265
5	1.319649076
10	1.035374328
20	0.6344081369
50	0.5068374597
100	0.3485700111

the *distribution* of those sums (and therefore the distribution of means) approaches the Normal, or Gaussian, distribution as the number of values included in the sum increases, in accordance with the central limit theorem (review Chapter 8).

For the purpose of performing hypothesis tests (here we go again!), we noted that what we need to do is determine the probability of finding the measured value of a given statistic under some null hypothesis. If the probability is high, we can accept the null hypothesis; if the value is improbable, we reject the null hypothesis.

So, if our statistic is an average, then by the central limit theorem we can find the probability levels to use from tables of the Normal probability distribution. We simply compute the number of standard deviations that our measured value of $\bar{X}$ is away from the hypothesized value of μ:

$$Z = \frac{\bar{X} - \mu}{\sigma / n^{1/2}} \tag{11-11}$$

Note that in the denominator of Equation 11-11, we must use the value for standard deviation *of the averages*, rather than the standard deviation of the population.

The symbol Z is, by convention, the symbol used to designate the number of standard deviations away from the hypothesized population mean.

The Big ªCatchº

Anyone who has read the whole book up to this point knows by now that whenever we come up with something that appears too simple, we usually have something up our sleeve. This time is no exception. What we have up our sleeve now is a big catch in the use of statistics, and in the use of the Normal distribution and in the use of the central limit theorem.

The existence of this catch is a forewarning of one of the pitfalls in the use of statistics that we are continually making snide asides about. In fact, while the Normal distribution is the most common distribution in nature, and the central limit theorem is the most important theoretical point in the science of statistics, Equation 11-11 is almost completely useless in practice, and so is the Normal distribution. The reason for this becomes clear when we examine the denominator of Equation 11-11 more closely. (No, forget the $n^{1/2}$—that is correct; you are not going to catch us out on the algebra.) Notice that the denominator contains the population standard deviation σ, rather than the sample standard deviation S. The problem is the σ. The use of Equation 11-11 is predicated on the fact that, somehow, we know the population standard deviation σ. How often do we know that, when we usually do not even know μ, the population mean? Not too often. But in order to use Equation 11-11, we must know σ. So what do we do?

Well, if we do not know σ, we can still calculate S, the sample standard deviation, and use S in place of σ, and $S/n^{1/2}$ instead of $\sigma/n^{1/2}$.

$$\frac{\bar{X} - \mu}{S/n^{1/2}} \tag{11-12}$$

If we do this, we should get something pretty close to what we would have found if we had used Equation 11-11, since we have already seen that S estimates σ.

There is only one small problem. If we look back at Chapter 5 we find, in Figure 5.2, a small diagram of the distribution of standard deviations. While the histogram (distribution plot) in that figure is for one particular case, it shares the characteristics of virtually all distributions of standard deviations. In particular, it is weighted toward smaller values; it contains values that go almost down to zero (for continuous distributions it would extend to zero), and it has a very long tail at the high end.

Since the distribution of S, the sample standard deviation is weighted toward smaller values, and Equation 11-12 uses S in the denominator, it is reasonable to expect that the quotient will be weighted toward *larger* values.

Back to the Computer

Table 11-2 presents a BASIC program that, as is our wont, takes our favorite population (the integers 1–10) and computes statistics from it. The program is a modification of the ones used previously, so if you have been following along and performing the exercises we have presented, you have a head start. In this case, we compute the values that result from applying Equation 11-12 to samples from the population. As we will recall from our previous usage, the purpose of the program is to compute an exhaustive list of the possible values of the statistic under study, by computing the statistic on all possible samples that can be drawn from the population. This creates what is, in effect, a population of the possible values of the statistic. Then, drawing a random sample from the population of possible values of the statistic is equivalent to computing that statistic on a random sample drawn from the original population, in this case, the population of the first 10 integers. The difference is that, by having the population of values of the statistic, it is easy to determine the fraction of those values lying within any given range, or above/below any specified value, which is then equivalent to determining the probability of obtaining those values when the statistic is calculated from a random sample obtained from the original population.

TABLE 11-2
Program listing used to generate tables.

```
1 N1 = 5 : REM set N1 equal to number of values averaged
10 DIM POP(10),DAT(N1),PERCENT(11),ALLMEAN(5000),
   COUNTMEAN(5000)
16 COUNTZERO = 0:MEANTOT = 0 : REM number of values saved
20 FOR I = 1 TO 1000: ALLMEAN(I) = 0:NEXT I: REM initialize
30 FOR I = 1 TO 10: READ POP(I): NEXT I: REM get data
   (from line 4500)
35 FOR I = 1 TO 11:READ PERCENT(I):NEXT I
40 FOR I = 1 TO 10 : REM set up all combinations of 6
42 FOR J = 1 TO 10
43 FOR K = 1 TO 10 : REM delete lines 46 - 43 successively for
44 FOR L = 1 TO 10 : REM fewer variables
45 FOR M = 1 TO 10
50 GOSUB 9000 :REM get mean/s.d. of N1 numbers
55 REM PRINT XBAR,S1 :REM diagnostic print
57 REM
58 REM See if values already in list:
59 REM
75 FOR L1 = 1 TO MEANTOT: IF XBAR = ALLMEAN(L1) THEN GOTO 95
80 NEXT L1
85 COUNTMEAN(MEANTOT + 1) = 1 : ALLMEAN(MEANTOT + 1) = XBAR:
   REM not in list
90 MEANTOT = MEANTOT + 1 : GOTO 100: REM add and update list
95 COUNTMEAN(L1) = COUNTMEAN(L1) + 1 : REM in list;
   increment count
100 REM
330 NEXT M : REM runs with fewer variables
340 NEXT L
350 NEXT K
360 NEXT J
370 NEXT I
390 PRINT MEANTOT : PRINT
400 M = MEANTOT
410 FOR I = 1 TO M
420 PRINT ALLMEAN(I),COUNTMEAN(I)
430 NEXT I
435 REM
440 REM sort the list of standard deviations
445 REM
450 FOR I = 1 TO MEANTOT-1
460 FOR J = I + 1 TO MEANTOT
470 IF ALLMEAN(I) =< ALLMEAN (J) THEN GOTO 500
480 TEMP = ALLMEAN(I):ALLMEAN(I) = ALLMEAN(J):
    ALLMEAN(J) = TEMP
490 TEMP = COUNTMEAN(I)
491 COUNTMEAN(I) = COUNTMEAN(J)
```

Table 11-2 *(continued).*

```
492 COUNTMEAN(J) = TEMP
500 NEXT J : NEXT I
510 REM
520 REM list sorted s.d.'s
530 REM
535 PRINT:PRINT "Sorted list:":PRINT
540 FOR I = 1 TO MEANTOT :PRINT ALLMEAN(I),COUNTMEAN(I) : NEXT I
550 REM
560 REM find the critical values
570 REM
575 PRINT:PRINT "CRITICAL VALUES FOR N = ";N1:PRINT
580 FOR I = 1 TO 11
590 TOT = (10^N1)-COUNTZERO: REM account for the instances
    when S1 = 0
600 CRITVALUE = PERCENT (I)*TOT:SUM = 0
610 FOR J = 1 TO MEANTOT
620 SUM = SUM + COUNTMEAN(J)
630 IF SUM = >CRITVALUE THEN PRINT PERCENT(I), ALLMEAN(J):
    GOTO 650
640 NEXT J
650 NEXT I
660 END
4500 DATA 1,2,3,4,5,6,7,8,9,10
4510 DATA.01,.025,.05,.1,.3,.5,.7,.9,.95,.975,.99
9000 REM
9001 REM compute means and s.d.'s from the population in array DAT
9002 REM
9050 SUMX = 0
9055 REM
9056 REM - Modify lines 9060 and 9070 appropriately for runs with
9057 REM fewer than 6 variables
9060 DAT(1) = POP(I):DAT(2) = POP(J):DAT(3) = POP(K)
9070 DAT(4) = POP(L):DAT(5) = POP(M)
9080 FOR N2 = 1 TO N1:SUMX = SUMX + DAT(N2):NEXT N2
9140 XBAR = SUMX/N1: REM sample mean
9150 SUMSQUARE = 0 :REM initialize
9160 FOR M1 = 1 TO N1: REM sum of squares
9161 SUMSQUARE = SUMSQUARE + (DAT(M1)-XBAR)^2 : REM sum of
     squares
9163 NEXT M1
9170 S1 = (SUMSQUARE/(N1-1))^.5 : REM s.d. by (n - 1) computation
9175 IF S1 = 0 THEN GOTO 9200:REM insert value instead of error
9180 XBAR = (XBAR-5.5)/(S1/SQR(N1)) :REM 5.5 = POP MEAN
9190 RETURN
9200 COUNTZERO = COUNTZERO + 1 : XBAR = 1000:RETURN : REM dummy
     value
```

Since our interest here is in determining the distribution of results, the program in Table 11-2 is set up to compute

$$\frac{\bar{X} - \mu}{S/n^{1/2}}$$

using all possible samples of size five from the population. It also does some other things, that we will discuss below.

A few words about the program. First of all, it can be easily changed to accommodate corresponding computations for samples of less than five. Only four program statements need to be modified.

To make such modifications, make the following changes: statement 1, on the first line, should reflect the number of samples that are being averaged in any run. Thus, to change the program to compute the statistics for samples of size three, for example, statement 1 should read:

1 N1 = 3

Program statements 40–45 set up the combinations; to change the number of variables, these lines must be deleted successively, starting from statement 45. Thus, to continue our example, statements 45 and 44 should be deleted in order to use samples of size three.

Statements 330–370 complete the loops set up by statements 40–45, therefore when any of the statements 40–45 are deleted, the corresponding statements from the group 330–370 must also go. In our example, statements 330 and 340 should be deleted.

Finally, statements 9060 and/or 9070 must be changed so as to include only those variables that are active. To change to three variables, as we are doing in our example, statement 9070 should be deleted entirely.

Now, some discussion of what the program does and how it does it. Statements 40–370 set up nested loops that ensure all possible combinations of random samples are taken from the population.

At statement 50 we get a value corresponding to the combination selected, and in statements 75–95 we create two lists. One list contains every unique value created during the course of the computations. The second list contains a summary of how many times each unique value appeared.

After printing the lists, statements 440–500 sort the lists, in ascending order of the values. This is in preparation for the real job of the program, which is to determine the percentage points, in order to calculate the critical values corresponding to the various critical levels.

After printing the sorted list, statements 560–660 determine the values that are just above known fractions of the number of results. Recalling our discussion of probability, these fractions represent the probability that a randomly chosen combination, which is equivalent to the probability that a randomly chosen value of $(\bar{X} - \mu)/(S/n^{1/2})$ will be above or below the demarcation point.

In running the program, it is best to make several versions, one each for different size samples to be averaged. Cover the range 2–5, and start with the smaller sets. The amount of computation required increases very rapidly with the sample size, and you may not want to tie up your computer for as long as it takes to do the larger runs. In fact, we cheated when we ran the programs. We did not actually run the program listed, except for sample sizes of two and three.

To be honest, the programs we actually ran were modifications of the ones listed, modified to run faster. In fact, we modified them all the way to FORTRAN. The BASIC program listed, for samples of size five, would require several days to run; the FORTRAN version ran in a few hours.

We also made another modification, that we used for samples of size six and larger. Since the number of combinations of n samples in an exhaustive list equal 10^n, the exhaustive listing of all the combinations of six or more samples from the population of integers would have been too time-consuming, even in FORTRAN. We therefore used one of the other techniques that we have discussed for generating probability tables (see Chapter 6): a Monte Carlo method. In this modification, the computer selected a sample of size six (or whatever size was under study, (n)) by selecting n integers at random, and computing the statistics of that set. The selection and computation process was then repeated 200,000 times for sets of six integers, and 300,000 for the larger sets. The probability points were then determined from the set of 200,000/300,000 statistics which, though not exhaustive, is very representative indeed.

In Chapter 12 we will examine the results of running the program. We recommend that you, our readers, also enter and run the program, and examine the results, at least for sample sizes of two and three. That way you will be caught up to us and ready to continue.

12

THE T STATISTIC

We are ready to continue the discussion we began in the previous chapter. Actually, this chapter is really a direct continuation of the previous one, so we strongly recommend that you go back and review it. Having done that, we are now ready to go on, and examine the results of running the program we created to calculate the percentage points of the distribution of the statistic $(\bar{X} - \mu)/S$.

Now for the results. Table 12-1 presents a list of values obtained from running the program. In fact, Table 12-1 is the sorted list printed by statement 540. Table 12-1 is the result of running the program for samples of size two, i.e., each average is the result of averaging two samples from the population of integers.

What do we find in Table 12-1? Well, for all combinations of two integers there are 100 total values generated; of these, 10 were rejected because S was zero, leaving 90 results to appear in the table. These 90 results are spread out over 37 unique values. The number of unique values of the statistic we are studying, for several different values of n, are presented in Table 12-2. The number of unique values is a much larger number than that of either the unique means or unique standard deviations alone (for samples of three, compare Table 12-1 with Table 5-1. This continues to hold true for sets of different sizes. Because there are so many unique values, there are not many entries corresponding to any given value. Thus, "eyeballing" Table 12-1 does not give a good feel for what is happening; most of the values have a count of two. On the other hand, we can do the following: since almost all the values have the same count, we can eyeball

TABLE 12-1
List of values of $(\bar{X} - \mu)/S$ for samples of 2.

Value	Number	Value	Number	Value	Number
−8.0000000	2	−0.5000000	4	0.5000000	4
−6.0000000	2	−0.4000000	2	0.6666667	2
−4.0000000	2	−0.2857143	2	0.7500000	2
−3.5000000	2	−0.2500000	2	0.8000000	2
−2.5000000	2	−0.1666667	2	1.2500000	2
−2.0000000	4	−0.1250000	2	1.3333334	2
−1.5000000	2	0.0000000	10	1.5000000	2
−1.3333334	2	0.1250000	2	2.0000000	4
−1.2500000	2	0.1666667	2	2.5000000	2
−0.8000000	2	0.2500000	2	3.5000000	2
−0.7500000	2	0.2857143	2	4.0000000	2
−0.6666667	2	0.4000000	2	6.0000000	2
				8.0000000	2

their spread. Doing this, we notice that the differences between adjacent values are smaller the closer the values are to zero. This is the indication that, for the distribution of $(\bar{X} - \mu)/S$, the points are densely packed around zero and sparse away from zero. Thus, we see that this statistic that we are investigating also tends to cluster around zero, and the density of the values decreases as we move away from zero, just like statistics following the Normal distribution.

The curve of probability density for the Normal distribution is often called "the bell-shaped curve". This is a misnomer. As we see, the density of $(\bar{X} - \mu)/S$ is also bell-shaped; thus, the curve corresponding to the Normal distribution is only

TABLE 12-2
Characteristics of the distributions of $(\bar{X} - \mu)/S$ for various sample sizes.

Sample size	Number of unique values	Range of values
2	37	−8 to 8
3	217	−12.5 to 12.5
4	463	−17 to 17
5	1431	−21.5 to 21.5
6*	2103	−26 to 26
7*	3378	−30.5 to 18.5
8*	2264	−17.08 to 21.18
10*	5784	−20.05 to 14.99
15*	10331	−9.22 to 8.59

*Values obtained from Monte Carlo calculation instead of exhaustive compilation.

one of many bell-shaped curves (we point out, incidentally, that the binomial distribution also follows a "bell-shaped curve").

Perhaps, however, the statistic we are investigating is also Normally distributed, we simply have not tested it for Normality. This is not the case. The verification of this fact can be seen in Table 12-3. Table 12-3 lists the values of standard deviation for which the corresponding fractional number of points is less than that value. For example, for averages of three samples, 5% (0.05) of the values were less than -3.499 (i.e., more than 3.499 standard deviations away from μ in the negative direction). Similarly, for averages of five samples, 90% of the values were less than 1.511.

Table 12-3 is, in fact, the statistical table corresponding to the distribution of $(\bar{X} - \mu)/S$, for several values of n. As such it is the exact analog of statistical tables for other more standard statistics, such as are found in the usual compilations of statistical tables.

We will recall from Chapter 6 that the Normal distribution is distributed so that the percentage points fall at fixed values of standard deviation. In Table 12-3 we see two important differences between the distribution of $(\bar{X} - \mu)/S$ and the Normal distribution. First, the various percentage points fall at a different number of standard deviations, depending upon the value of n, the number of samples that were included in each mean.

The second important difference from the Normal distribution is that the number of standard deviations away from the population mean corresponding to a given percentage point decreases as n increases. We noted from the previous chapter that, because the distribution of S is weighted toward smaller values, the distribution of $(\bar{X} - \mu)/S$ should be weighted toward larger values, and so it is. This distribution tends to have a very long tail, with many more points at high values of standard deviation than the Normal distribution does; furthermore, the number of points at high values of S.D. increases as n decreases. This leads us to an important characteristic of the statistic we are studying: the fewer the samples contributing to the mean, the more widely distributed are the resulting values of the statistic, even in comparison with the standard deviation. This is above and beyond the tendency of means to congregate around μ for large samples and spread away from μ for small samples; for in that case, for which the Normal distribution applies, at least the percentage points are fixed.

We have generated these results from one particular, and very limited population: the uniformly distributed population of the integers 1–10. However, similar results would have been obtained if the initial population itself were Normally distributed. We state this fact without proof. Formal proofs are available, but are even more difficult and complicated than the proof of the central limit theorem, so we will not even discuss the proof. However, anyone who wishes to verify the fact can modify the BASIC program so that the variable POP contains a normally distributed population of numbers

TABLE 12-3

Percentage points for $(\bar{X} - \mu)/S$ of averages of n values taken from the population of the integers 1–10. Each entry represents the number of standard deviations for which there are a corresponding fractional number of points less than that value.

Fractional points (probability)	n								
	2	3	4	5	6*	7*	8*	10*	15*
0.010	−8.000	−8.500	−6.062	−4.570	−3.884	−3.517	−3.307	−2.987	−2.713
0.025	−6.000	−5.500	−3.666	−3.129	−2.782	−2.593	−2.485	−2.337	−2.184
0.050	−4.000	−3.499	−2.634	−2.255	−2.078	−1.983	−1.914	−1.844	−1.769
0.100	−2.500	−2.165	−1.666	−1.511	−1.453	−1.425	−1.403	−1.377	−1.337
0.300	−0.500	−0.500	−0.541	−0.532	−0.533	−0.533	−0.529	−0.530	−0.528
0.500	0.000	−0.064	0.000	−0.049	0.000	0.041	0.000	0.000	0.034
0.700	0.500	0.500	0.541	0.532	0.533	0.525	0.529	0.531	0.528
0.900	2.500	2.078	1.666	1.511	1.459	1.425	1.403	1.369	1.338
0.950	4.000	3.499	2.634	2.255	2.088	1.983	1.914	1.844	1.764
0.975	6.000	5.500	3.666	3.129	2.800	2.593	2.457	2.321	2.190
0.990	8.000	8.499	6.062	4.570	3.927	3.577	3.307	3.009	2.719

*Values obtained from Monte Carlo calculation instead of exhaustive compilation.

(expand POP to contain at least 100 values) and then run that. The fact that similar results would come from a Normally distributed population is the hook upon which we can hang a generalization to all populations: The averages of samples selected from the Normal distribution are also Normally distributed. However, by the central limit theorem, so are the averages of samples from *any* population. Therefore, the behavior of averages of samples from any population will have the same behavior as averages of samples from the Normal distribution, and if this behavior is the same as from our uniformly distributed population, then so is the behavior of averages from any population.

Note, by the way, that this is the true importance of the central limit theorem: it does not usually apply directly to measurements we make or data we collect, but having proven something for the Normal distribution, the central limit theorem allows us to generalize those results to other populations.

What, then, is the distribution corresponding to $(\bar{X} - \mu)/S$? It turns out that this is the distribution known as Student's t. This distribution can be described by a mathematical formula, and we present the formula for the sake of completeness (its derivation can be found on page 237 in Reference (12-1))

$$t(t,f) = \frac{\Gamma\left(\frac{f+1}{2}\right)}{(\pi f)^{1/2}\Gamma\left(\frac{f}{2}\right)} \int_{-\infty}^{t} \left(1 + \frac{x^2}{f}\right)^{-(f+1)/2} \mathrm{d}x \qquad (12\text{-}1)$$

This formula describes a continuous function, rather than the discrete one that we generated Table 12-3 from, but comparison of published t-tables, which are computed using Equation (12-1), with Table 12-3 reveals reasonably good agreement between the two sets of values. In making such a comparison, an entry in Table 12-3, which is compiled based on n, the number of samples contained in each average, should be compared with the t-table entry for $n - 1$, the number of degrees of freedom, which is more commonly used for those compilations.

The Nature of the *t* Distribution

What is the nature of the t distribution? As we surmised above, this distribution can also be represented by a symmetric, bell-shaped curve. The shape of the curve changes as the number of degrees of freedom change; for fewer degrees of freedom there are much higher percentages of the values at large values of t. This is a consequence of something we noticed in the previous chapter: that the distribution of standard deviations is skewed toward smaller values, therefore, the distribution of values obtained by *dividing* by the standard deviation tends to have a bigger share of large values.

A characteristic of t that is not too clear in Table 12-3, but can be seen in the published tables if you look for it, is that for very large numbers of degrees of freedom the t-distribution starts to approximate the Normal distribution. When doing hypothesis testing, if you are working with large numbers of samples, and the values you want are beyond the limits of the t-table that is available, then a table of the Normal distribution can be used to determine critical values for the hypothesis test.

How and when do we use the t distribution? We use it the same way we use the Normal distribution, to determine confidence levels, confidence limits, and critical values for performing hypothesis tests. The difference is that to use the Normal distribution, the population value of standard deviation, σ, must be known. If the population value is unknown, and we must use the sample standard deviation S, then the t distribution must be used to determine the confidence limits and critical values.

$$Z = \frac{\bar{X} - \mu}{\sigma / n^{1/2}} \tag{12-2A}$$

$$t = \frac{\bar{X} - \mu}{S / n^{1/2}} \tag{12-2B}$$

The symbol t is usually not used alone. The conventions of statistical terminology dictate that when a value is to be specified for t, both the confidence level and the number of degrees of freedom are also to be indicated; it is conventionally done by rewriting Equation (12-2B) as follows:

$$t(P, n) = \frac{\bar{X} - \mu}{S / n^{1/2}} \tag{12-3}$$

where P represents the confidence level defined by the α probability P for that value of t, and n represents the number of degrees of freedom.

Examples

We will end this chapter with some illustrative exercises.

Example 1: We would like to determine as best we can, the absorbance of a reference standard at a particular wavelength. Suppose we take five measurements of the absorbance at a wavelength and find an average $A = 0.78205$ with a standard deviation of 0.00018. The 95% confidence interval for the true value of A of this sample using this instrument at this wavelength is given by:

$$\mu = X \pm t(0.95, 4) \times (S/n^{1/2}) \tag{12-4}$$

where X is our estimated mean absorbance, t is the value of t from the t-distribution table, S is our sample standard deviation, and n is the number of measurements. Thus, for our specific example, the real estimate for A is

$0.78205 \pm 2.57 \times 0.00018/5$, and our spectroscopic measurement error is $\pm t \times S = \pm 2.57 \times 0.00008 = \pm 0.00021$. From this information we can specify that any single measurement that falls outside of the range of our 95% confidence interval $0.78205 \pm 0.00021 = 0.78184$ to 0.78226, would indicate a systematic change in the instrument readings (or perhaps a real change in the sample).

Example 2: We would like to determine how many measurements or spectra we would require to achieve a specific precision and accuracy for these measurements. We note that we define precision as the upper and lower confidence limit for a specific confidence level (e.g. $\alpha = 0.05$). For any given number of measurements we can increase our precision only by reducing the probability of being correct, while we can increase the probability of being correct if we reduce our precision requirements. If we wish to increase both our reliability (probability of being correct) and our precision (tightening of our confidence interval) we must increase the number of observations (n). The relationship between an increase in n and a decrease in standard error is due to the fact that n appears in the denominator of the standard error expression. This of course indicates that standard error is inversely proportional to sample size.
So then, how do we determine the number of measurements required to meet our individual requirements for precision and confidence? To illustrate this situation, let us use the following spectroscopic example. We repacked aliquots of a single specimen five times and obtained the five corresponding spectra for these samples. From these data we apply a quantitative algorithm and find our predicted analyte concentration (the mean of the five replicates) is 10.00% with a standard deviation (using $n-1$ degrees of freedom) of 1.00%. We have established that our analytical requirements for this analysis would ideally be 0.1% absolute (1 sigma), but we would accept 0.3% if we had to. Thus, the question is asked as to how many measurements (repacks) must we take to achieve the desired accuracy.
Letting S.D. stand for the standard deviation representing the desired accuracy of averaging five repacks, then we use the formula:

$$\text{S.D.} = S/n^{1/2} \tag{12-5}$$

For our example, we have $\text{S.D.} = 1.0/\sqrt{5} = 0.447$. Now we state those values that we know: our mean estimate $(\bar{X}) = 10.00$; $n = 5$; $S = 1.00$; and $\text{S.D.} = 0.447$. Our confidence limit is set at 0.95 and therefore our t-distribution value $t(0.95,4)$ from our table is 2.571. Thus, as $X = t \times \text{S.D.} = 10.00 \pm 2.571 \times 0.447 = 1.149$, we would have a "95% confidence" that our actual analyte concentration is between 8.851 and 11.149—not too impressive!

If we wish to reduce our error to 0.10% absolute (at 1 standard deviation), while maintaining our confidence limit, how many measurements would be required? Substituting our desired value of S for our previously calculated values, we have $0.10/1.149 = 1.00/\sqrt{n}$, thus $n = (1.00/0.0870)^2$, and the required n is approximately 132—not too practical!

Well then, what accuracy at 95% confidence could we expect with a reasonable number of spectral measurements (say, about 16)? To solve we use $\text{S.D.} = S/\sqrt{n}$ and therefore $\text{S.D.} = 1.149/4 = 0.287\%$. And so if our error is completely random and normally distributed (without a significant systematic component) we can calculate useful information related to accuracy, precision, confidence limits, and number of measurements required.

Example 3: Here we use real data that we have already presented: the transmittances corresponding to a 100% line that we presented in Chapter 6. We ask the question: is there any evidence for unbalance of the instrument in that data?

We have already presented most of the pertinent facts about that data; we review them here, and make sure that we include everything we need:

$n = 98$

$\bar{X} = 0.998$

$S = 0.0366$ (if this seems high, remember that the instrument performance was deliberately degraded).

We expect that a 100% line will have a theoretical value of unity; thus, this represents the value of μ that we will hypothesize and test against:

$$\text{Ho}: \ \mu = 1$$

We are concerned about the possibility that there is a systematic error, so that the data is from a population that does not conform to the null hypothesis Ho. This is written formally as an alternative hypothesis, Ha:

$$\text{Ha}: \ \mu \neq 1$$

Our test statistic is:

$$t = \frac{\bar{X} - \mu}{S/n^{1/2}}$$

performing the calculations:

$$t = \frac{0.998 - 1}{0.0366/98^{1/2}}$$

$$t = -0.002/0.00369 = -0.54$$

To perform the hypothesis test we must compare our calculated value of t with the critical values of t that we find in the tables. We find that, for our data:

$$-1.98 < t(\text{crit}) = t(0.95, 97) < 1.98$$

Thus, unless the calculated t is outside the range -1.98 to $+1.98$, we have no evidence that the instrument is unbalanced. Since the t we actually found, -0.54, is not outside that critical interval, we accept the null hypothesis and conclude that the instrument is not unbalanced; the measured difference of the 100% line from unity is explainable purely by the effects of the random chance variations due to the noise of the data alone.

A few words about our hypothesis test: first, note that while we calculated the divisor of Equation 12-3 we took the square root of 98, but when we compared the calculated t to the critical t we used the value 97. This is because Equation 12-3 specifies n, the number of samples, while performing the hypothesis test requires using the number of degrees of freedom. Since we calculated the sample mean and subtracted it from each data point, the data used to calculate S contains only $n - 1$, or 97 degrees of freedom.

Second, if you check a t-table you will find the critical t we used, 1.98, listed as the value of t corresponding to the 0.975 probability level, rather than the 0.95 level that we claim we are performing the hypothesis test at. Why this disparity? Recall that the 0.975 figure means that 97.5% of the points are below the corresponding value of t (1.98 in this case) and 2.5% are above $t = 1.98$. Remember further that the t-distribution is symmetric. Therefore, 2.5% of the points also lie below $t = -1.98$. Thus, by including points on both sides of the distribution in the rejection regions, rejecting 2.5% of the points on each side leaves 95% of the points in the acceptance region. If we had used the t value listed for 95% (1.66) we would have been testing at only a 90% critical level, because we would have included 5% on each side of the distribution, for a total of 10% in the rejection region.

References

12-1. Cramer, H., *Mathematical Methods of Statistics*, Princeton University Press, Princeton, NJ, 1966.

13

Distribution of Means

An Explanation

Well, we have spent the last couple of chapters doing a lot of algebra and a lot of computing, and now it is time to step back a bit and see what it all means. For example, in Chapter 11 we derived the well-known result:

$$\text{S.D.}(\bar{X}) = \text{S.D.}(X)/n^{1/2} \tag{13-1}$$

This equation states that, if we have available a quantity of numbers, X_i, where the quantity $= n$, and each number represents a certain population value (μ) with a random value added to μ, then these numbers will have a certain standard deviation, which we represent with the symbolism S.D.(X). The equation further states that if we take the average of these n numbers, the average will have a standard deviation of S.D.(X)/$n^{1/2}$.

Now, the standard deviation of the numbers X has a pretty straightforward meaning, because we can apply the formula for calculating standard deviations (see, for example, Equation 11-1) to the several values of X_i. For the standard deviation of the average, $\bar{X}$, on the other hand, we have the formalistic calculation that Equation 13-1 specifies, but what does this signify? The average is only a single number; is it meaningful to talk about the standard deviation of a single number?

Let us approach this question by reviewing something we did along the way. In Table 11-1 we showed the results of computing many means, and taking the standard deviations of those sets of means. We found that, indeed, the standard deviations of the means followed Equation 13-1. Well, that is fine when we have many means, but still leaves us with the problem of understanding the meaning of the standard deviation of only a single mean.

For example, the single mean could be any one of the values in Table 11-1; if we had computed only one of those means instead of all of them, it would have been a random sample of the distribution of means. Equation 13-1 states that this sample of one mean has the same standard deviation as the sample of 20 means (for the corresponding n for each, of course) that occurs in each part of that table. Suppose we try this out, by taking the formula for standard deviation:

$$S = \left(\frac{\sum_{i=1}^{n} (\bar{X} - \bar{\bar{X}})^2}{n - 1} \right)^{1/2} \tag{13-2}$$

where $\bar{X}$ represents a given single mean value, and $\bar{\bar{X}}$ represents the mean of all the means. Now let us apply this equation to the single value:

$$S = \left(\frac{\sum_{i=1}^{n} (\bar{X} - \bar{\bar{X}})^2}{1 - 1} \right)^{1/2} \tag{13-3}$$

Since there is only the single value available, $\bar{X} = \bar{\bar{X}}$, and Equation 13-3 reduces to the classic form of an indeterminate value:

$$S = 0/0 \tag{13-4}$$

Equation 13-4 is interesting enough in its own right to be worth a digression here. What does *this* mean? We have noted several times previously that as more and more data are included in the calculation of any statistic, the statistic becomes better and better known, at least insofar as repetitive computations of that statistic cluster more and more closely around the population value for the corresponding parameter.

Now let us look at this picture the other way, with particular reference to the calculation of standard deviation. The calculation of standard deviation involves the ratio of two other quantities, the sum of squares, which goes into the numerator and the degrees of freedom, which goes into the denominator. As more and more data are included in the standard deviation formula, both of these quantities increase numerically, as well as becoming better and better defined statistically. Conversely, as fewer and fewer data points are included, both quantities decrease, and become more and more poorly defined. As the number of data gets less, our certainty as to the value of the standard deviation becomes lower and lower. Furthermore, the magnitudes of both quantities becomes smaller and smaller, reflecting the fact that each number contains less and less information. At the limit, when we have only one datum, we run out of sums of squares and degrees of freedom at the same time: both numbers have decreased to zero, resulting in a mathematically indeterminate form. More to the point, if we have no computable difference, in actuality we really have no information whatsoever regarding the variability of the data (not even one degree of freedom's worth!), so that the standard deviation of the result is, in fact, undetermined. In this sense, the mathematical indeterminacy is a correct indication of the real indeterminacy of the result.

The same argument holds when, rather than using means, we use the original data; when we run out of sums of squares and degrees of freedom at the same time, the standard deviation is indeterminate, both in the mathematical sense and in the intuitive sense: we have absolutely no information concerning its value.

We may also note here that using the formula for standard deviation with n in the denominator would lead to a most incorrect and inappropriate result: if we substitute a single number into that formula, we come to the conclusion that the standard deviation $= (\bar{X} - \bar{\bar{X}})/n^{1/2} = (0/1)^{1/2} = 0$. Well, we know that *that* is certainly not so; data that exhibit randomness must have a non-zero standard deviation. Therefore, we must conclude that the other form ($n - 1$ in the denominator) at least makes more sense (whether or not it is more useful) but we are still left in a quandary.

Back to our main line of discussion. In our digression we seem to have demonstrated beyond all doubt that, in fact, a single number does not have a defined value of standard deviation at all, let alone one that has meaning. We really seem to have dug a hole deeper and deeper that we now have to climb out of to explain this whole mess.

We did this, of course, out of our own ulterior motives. In this case our motive was to give ourselves the opportunity to impress on our readers the importance of distinguishing between population parameters and sample statistics for the same nominal quantity. In Chapter 3 we pointed out that the quantities we label with the names "mean" and "standard deviation" (as well as every other statistic) each have different import depending upon whether the label is applied to a quantity derived from a population or from a sample. Here we have a situation where making the distinction and understanding the difference is not merely a matter of some vague and nebulous "correct" way of doing a computation, but is absolutely crucial to comprehending what is going on.

Trying to apply the standard deviation calculation to a single number resulted in Equation 13-4, an indeterminate form, because we were trying to compute the difference between our number $\bar{X}$ and the mean of the various $\bar{X}$, $\bar{\bar{X}}$, when the mathematics forces them to be exactly equal. On the other hand, all statistics are intended to estimate the corresponding population parameter. In this case, $\bar{\bar{X}}$

is the estimate of μ. If we compute the standard deviation using differences from μ, the problems go away. $\bar{X} - \mu$ has some non-zero value, and since there were no statistics computed prior to calculating that difference, no degrees of freedom were lost. Thus, this difference is a computation of S with one degree of freedom, and estimates σ with one degree of freedom.

Another way to look at it is to consider this value of $\bar{X}$ as a single random sample from the population of all possible values of $\bar{X}$—in other words, from the distribution. This view of the situation is important, because having recognized the distribution, we can note a sort of a corollary to the interpretation of probabilities. Ordinarily we interpret the distribution curve (of whatever distribution we happen to be working with) as the probability of obtaining a value at a given distance from the corresponding parameter for the given population. The corollary is to consider the central point of the distribution as representing a given measured value; the distribution then can be seen to represent the probability of finding the population parameter at that same distance.

This is the meaning of Equation 13-1: the standard deviation gives the width of the probability curve around the single measured (or calculated) value that defines where the true (or population) value can be found.

S.D.($\bar{X}$), then, is an estimate, based on the sample used, obtained from the original population of data, of the value of σ for the population of means. This seems a rather indirect way of computing the standard deviation of the population of means: why not simply compute it directly from $\bar{X}$? Answer: because we do not know μ. If we knew μ we could, indeed, do the direct computation. Without the knowledge, however, we are forced to use the indirect method.

Is our explanation, which is based on computing $\bar{X} - \mu$, valid if we do not know μ? The answer is: Yes. The distribution of values of $\bar{X}$ around μ exists even when we do not know μ, and has a standard deviation equal to S.D.$(X)/n^{1/2}$, and that is the only requirement, for it is this distribution that determines the probability levels. Knowing the distribution, we can then turn the argument around and state that, knowing $\bar{X}$, we can determine the probability of finding μ within any given interval around that computed value of $\bar{X}$.

Knowledge of μ (or lack thereof) affects only the type of computation that we can use to calculate S.D.($\bar{X}$).

There is another aspect to this: the indirect computation of S.D.($\bar{X}$) would be preferred over the direct computation even if μ were known. The reason is that if we compute $\bar{X} - \mu$, it estimates σ for the distribution of means with only one degree of freedom. On the other hand, estimating this quantity by dividing the standard deviation of the parent population by the square root of n results in an estimate with $n - 1$ degrees of freedom, the same reliability as is obtainable for the parent population.

An Explanation of the Explanation

So now we have gone through a long, somewhat rambling discussion to explain the meaning of the standard deviation of a single number. There are two reasons we spent all the time and effort on this discussion. One reason is the one stated above: to give ourselves the opportunity to emphasize the importance of distinguishing between populations and samples, and the import of the difference between computations based upon those two concepts.

The second reason was to bring out the fact that the means themselves do, in fact, have a distribution of their own. We have seen previously that, by virtue of the central limit theorem, this distribution will be Gaussian. We have now progressed further, and shown that this Gaussian distribution will have a standard deviation that can be related to the standard deviation of the parent population, and that can be determined from computations based on samples from the parent population. Except for pedagogical purposes, we do not need more than one sample from the distribution of means in order to be able to do useful things.

Figure 13.1 illustrates the relationship between the distribution of a population and the distribution of means taken from that population. For convenience we have assumed that the parent population is Normally distributed, although this

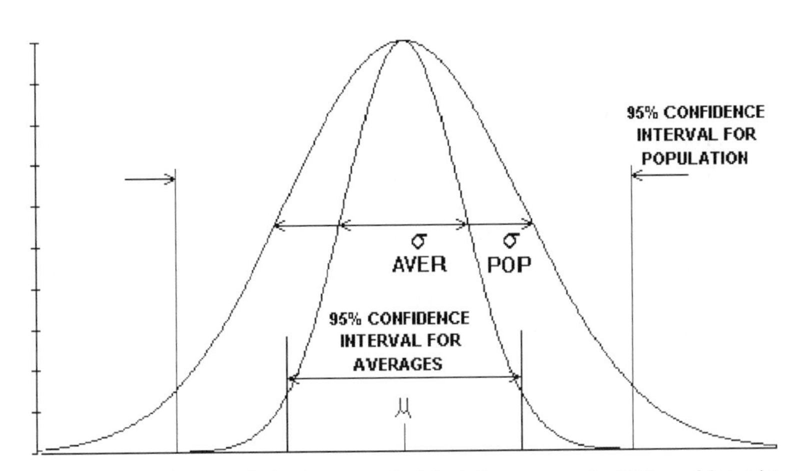

FIGURE. 13.1. A given population has a standard deviation = σ_{POP}; the 95% confidence interval surrounds μ at a distance that depends upon the distribution of the population; this distance is $1.96 \times \sigma_{POP}$ if the distribution is Normal. The averages of data taken from this population has their own value of σ; confidence limits for averages surround μ at $1.96\sigma_{AVER}$ since the averages are always Normally distributed.

need not be so. The distribution of averages, on the other hand, must be Normal, from the central limit theorem.

For both distributions, we have indicated the standard deviation, and the 95% confidence intervals. Since the distributions are Normal, each 95% confidence interval extends to 1.96 times the standard deviation of its corresponding distribution. That is an important point, because when we come to do hypothesis testing we will need to use the confidence interval for the appropriate distribution.

So, if we have data, and wish to determine if there is a shift compared to some standard value, we can use the mean of the data. If we do so, we can do our hypothesis testing by comparing our data not against the distribution of standard deviations of the data itself, but against the distribution of means of samples taken from that data. The advantage of doing this is that it provides a more sensitive test, because the standard deviation of the means is smaller than that of the data.

This is also illustrated in Figure 13.1. For the hypothetical data shown, the 95% confidence interval for the distribution of means is approximately the same as the standard deviation of the parent population. (Exercise for the readers: what value must n be in order for this to be exactly so?) If n were larger, the distribution of means would be even narrower, and differences could be detected even when they were less than the standard deviation of the data. This is an interesting point, because ordinarily something that is smaller than the standard deviation of the data would be considered "lost in the noise". However, we see that this is a bit of a misconception, because if enough data is available, even such small differences can be detected.

This point is illustrated in Figure 13.2. Here we have taken our favorite 100% line, the one we presented in Chapter 7, and offset the two halves from each other. Let us compare what we can see with what is really going on.

The standard deviation of the untouched data in this 100% line is 0.03776, and it contains 351 data points, 175 in each half. Therefore, the standard deviation of the average of the data in each half is, by Equation 13-1, $0.03776/175^{0.5} = 0.03776/13.22 = 0.00285$. Therefore, a deviation of more than $1.96 \times 0.00285 = 0.005586$ is more than randomness in the data can account for.

In the original data, the averages of the two halves of the data are 0.996486 and 1.000561. The differences of each of these values from unity are well within the tolerance that the randomness of the data permits. The difference between them, 0.00407, is also accounted for solely by the random nature of the data.

In the four sections of Figure 13.2, we have displaced the two halves of the data by adding a constant to the data in one half. The offsets added are 0.01, 0.02, 0.05 and 0.10, respectively; all of these are far more than can be accounted for by randomness alone. A hypothesis test against the null hypothesis that $\mu = 1$ would, of course, detect these added offsets since even the smallest of them

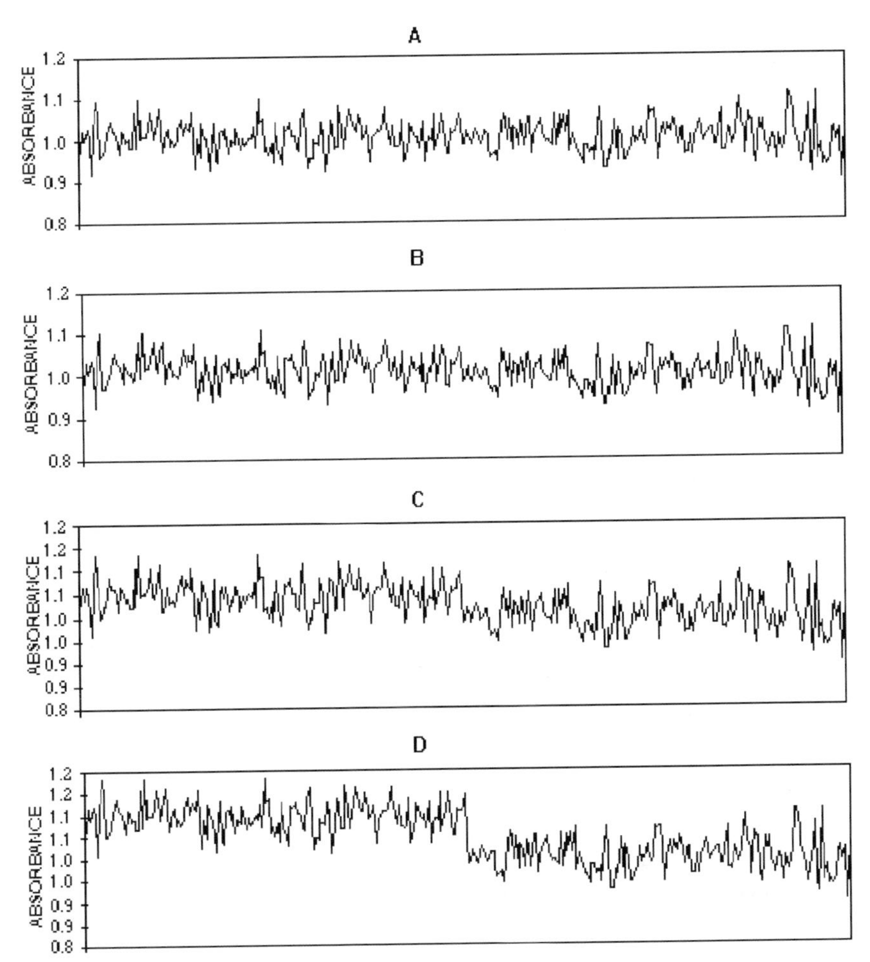

FIGURE. 13.2. When many data are available, very small offsets can be detected, much smaller ones than are noticeable to the eye. The random variation of the 100% line presented here is approximately 0.03. The four sections of this figure contain offsets of 0.01, 0.02, 0.05, 0.10 between the two halves of the spectrum. The offset is not noticeable to the eye until it becomes 0.05 or greater, as in section C, while a calculation would reveal the difference even in section A.

would result in the computed value of t that was statistically significant:

$$t = \frac{0.01}{0.03776/175^{1/2}} = 0.01/0.00285 = 3.51$$

Thus, the smallest offset is 3.51 standard deviations (of the distribution of means) and therefore statistically significant (as a reminder, "statistically significant" translates: "more than the randomness of the data alone can account for; something else is happening").

To the unaided eye, however, the 100% lines in Parts A and B of Figure 13.2, with offsets of 0.01 and 0.02, show no sign of this displacement. In Part C, the offset of 0.05 is marginally noticeable; looking at this, one might suspect that there might be an offset. Only in Part D is the fact of displacement really obvious. The 100% line in Part D corresponds to a displacement of 0.1; this value is three times the standard deviation of the parent population, which is the value the unaided eye compares the offset to.

Thus, the use of the average provides a sort of "magnifying glass" with which to look at data, and is a very powerful tool for performing hypothesis tests. It allows very small changes to be detected, even smaller than the noise level, if the data is appropriate.

There is another consequence of the difference between the use of population parameters and sample statistics. We have seen one example of this difference in the previous two chapters. The distribution of $(\bar{X} - \mu)/\sigma$ is not the same as the distribution of $(\bar{X} - \mu)/S$; the former is Normally distributed, while the latter follows a t distribution. Since we spent two chapters discussing the t distribution, we do not wish to belabor the point here, but do wish to point out that this phenomenon generalizes: the distribution of values obtained when sample statistics are used is not the same as when population parameters are used.

14

ONE- AND TWO-TAILED TESTS

This chapter is going to be rather dull and boring. The topic we will discuss is important, but our discussion of it will be rather mundane, nor do we have any cute tricks up our sleeves to enliven the presentation (at least, not at this point, when we are just starting to write it—maybe something will occur to us as we go along. We can always hope). So, onward we push.

Figure 14.1-A shows a diagram we should all be used to by now, a Normal distribution plot with the 95% confidence interval marked. Recall the meaning of this: given a random, normally distributed variable, 95% of the time a randomly selected value will fall between the lines, so that the probability of finding a value between $\mu - 1.96\sigma$ and $\mu + 1.96\sigma$ is 0.95. Since probabilities must add to 1.0, the probability of finding a value outside that range is only 0.05, so that if a measurement is made that does in fact fall outside that range, we consider that the probability of it happening purely because of the random variability of the data is so low that we say, "this value is too improbable to have happened by chance alone, something beyond the randomness of the reference population must be operating". This statement, or a variant of it, is the basis for performing statistical hypothesis testing. The Normal distribution, of course, applies only to certain statistics under the proper conditions; so far we have discussed the fact that it applies to data subjected to many independent sources of variation and to the mean of a set of data of any distribution.

At this point we will note but not pursue the fact that there are two possible explanations why data can fall outside the 95% confidence interval: either

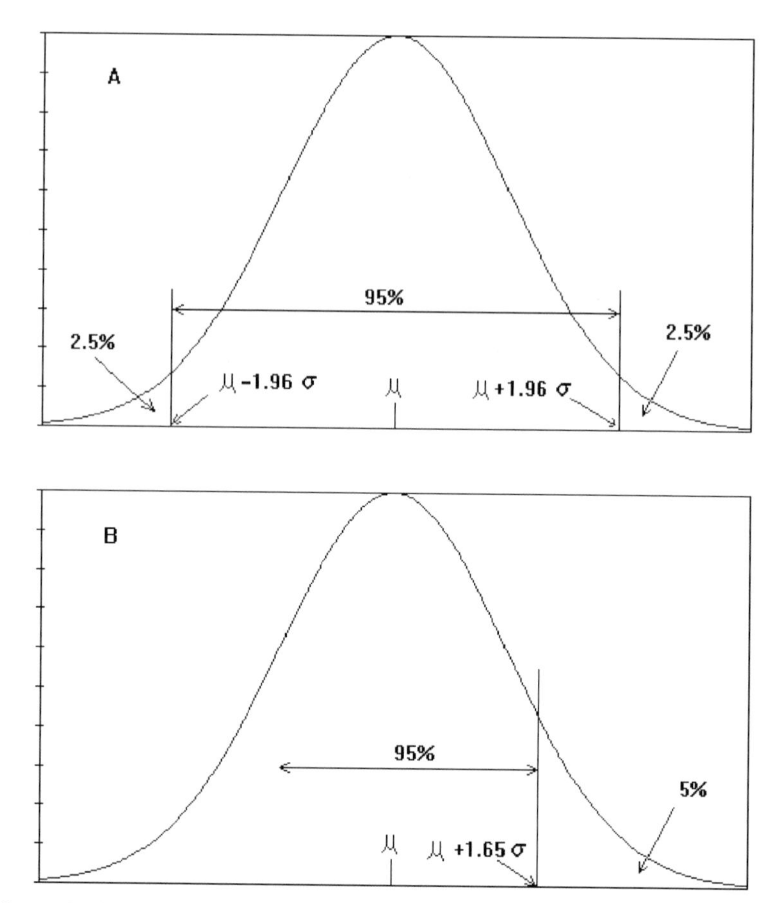

FIGURE. 14.1. Part A shows the conditions for a two-tailed hypothesis test. Each of the tails, which start at $\pm 1.96\sigma$, contains 2.5% of the possible cases; this is appropriate for testing if a measurement is different from the hypothesized value of μ. Part B shows the conditions for a one-tailed test; this is appropriate for testing whether a measurement is higher (or lower, if the other tail is used) than the hypothesized population value. For a one-tailed test, the tail starts at 1.65σ.

the population mean for the measured data is not the hypothesized mean, or the population value of standard deviation is not the hypothesized value. Either of these situations is a possible explanation for the failure of any given null hypothesis concerning μ.

What we do wish to point out here is that the 95% confidence interval shown in Figure 14.1-A is symmetrical, and is appropriate to a null hypothesis of μ being equal to some value that we wish to test, against the alternative hypothesis that μ equals some other value. It is this alternative hypothesis that is the subject of this chapter.

Stating the alternative hypothesis that way, i.e., simply allowing μ not to equal the hypothesized value, leads to the diagram of Figure 14.1-A and the associated probabilities, that 95% of the time a sample from the specified population falls within the indicated range, 2.5% of the time is above the range and 2.5% of the time is below the range.

Now you are probably sitting there wondering "What else could there be? μ either equals the hypothesized value or it doesn't." Well, that is almost but not quite true (surprise!). When we draw a diagram such as in Figure 14.1-A, we make an implicit assumption. The assumption is that, if μ does not equal the hypothesized value, the probability of it being above the hypothesized value is equal to the probability of it being below the hypothesized value.

This assumption is sometimes true, but sometimes not. If it is true, then Figure 14.1-A is a correct and proper description of the situation. On the other hand, suppose that assumption is not true; then we have a problem. To illustrate the problem, let us suppose that the physical situation prevents μ from ever being less than the hypothesized value; it can only be greater. In this case, we would *never* obtain a value below the low end of the confidence interval due to failure of the null hypothesis; and under these conditions the lower end of the distribution might as well not exist, since only the upper end is operative.

But random samples from the distribution obtained from the population specified by the null hypothesis will exceed the upper bound only 2.5% of the time under these conditions, not the 5% originally specified. Thus, under conditions where the possible values are restricted, what was supposed to be a hypothesis test at a 95% confidence level is actually being performed at a 97.5% confidence level.

Intuitively you might like the extra confidence, but that is not an objective criterion for doing scientific work. If a true 95% confidence level for the hypothesis test is desired, it is necessary to adjust the position of the upper confidence limit to the point where, in fact, 95% of the samples from the test population fall within it. In the case of the Normal distribution, this value is $\mu + 1.65\sigma$. This value compensates for the fact that the lower confidence limit has disappeared. If a higher confidence level is desired, it should be introduced explicitly, by setting the confidence level to 97.5% (or 99%, or whatever confidence level is appropriate).

Figure 14.1-B presents the probability diagram for this condition. Ninety-five percent of the data lies below $\mu + 1.65\sigma$ and 5% lie above. Contrast this with Figure 14.1-A where the corresponding points are $\mu \pm 1.96\sigma$. The region outside the confidence limits is, in both cases, called the rejection region for the hypothesis test, because if a measurement results in a value outside the confidence interval, the null hypothesis is rejected. In the case shown in Figure 14.1-A the rejection region includes both tails of the distribution; in

Figure 14.1-B the rejection region includes only one tail of the distribution. For this reason, the two types of hypothesis test are called two-tailed and one-tailed tests, respectively.

Two-tailed tests are used whenever the alternative hypothesis is that the population parameter is simply not equal to the hypothesized value, with no other specification. One-tailed tests are used when the alternative hypothesis is more specific, and states that the population parameter is greater (or less, as appropriate) than the hypothesized value. We have presented this discussion specifically in terms of the Normal distribution. That was useful, as it allowed us to show the situation for a particular distribution that was well known, and furthermore, allowed us to use actual numbers in the discussion. Notice, however, that the only item of real importance is the probability levels associated with given values of the distribution. The point, of course, is that the discussion is general, and applies to the t statistic, as well as all other statistics, including ones that we have not yet discussed.

We have noted above that, for the Normal distribution, and for the t distribution, the distributions are symmetric and the positions of the critical values for a two-tailed test are symmetrically positioned around the population parameter. We will find in the future that the distributions of some statistics are not symmetric; how do we handle this situation? The key to the answer to this question is to note that it is always the probability levels that are important in statistical testing. For asymmetric distributions we can still find values for which the probability of obtaining a reading equals a given, predefined percentage of all possible random samples. These then become the critical values for hypothesis testing. It is of no consequence if they are not uniformly positioned around the population parameter: that is not a fundamental characteristic of the critical values, merely an happenstance of the nature of the Normal and t distributions.

But, you are probably asking yourselves, when would someone want to make the more specific alternative hypothesis, rather than just making the simpler hypothesis of $\mu \neq$ (test value)? In textbooks of statistics the usual examples involve a "manufacturer" who manufactures (fill in the blank) and who wants to know if a new manufacturing procedure is "better" than the old one in some way. Clearly the "manufacturer" is not interested in a procedure that is worse.

However, as an example of a situation of more interest to us as chemists where the concept of one-tailed testing applies, let us consider the question of detection limits.

Considerable literature exists concerning the definition of the detection limit for a given analysis; two recent articles have appeared (in the same issue of *Analytical Chemistry*—interest must be growing!) which treat the subject from the chemist's point of view (14-1,14-2) and include a good list of citations to older literature.

We are not going to enter the fray and try to come up with a definition that will suit everybody. Not only are people too cantankerous in general for such an attempt to succeed, but also as we shall see, different definitions may be appropriate in different circumstances. What we will do, however, is to discuss the subject from the statistical point of view; we hope that the insight that you, our readers, gain, will allow you to formulate definitions to suit your needs at the appropriate time.

What does the concept of a detection limit entail? In general, it allows the analyst to make a statement concerning the sensitivity of an analytical procedure. Presumably, in the absence of the analyte of interest, the method (whether it be specifically spectroscopic, or more generally, instrumental, or most generally, any method at all) will give some value of a measured quantity, that can then be called the "zero" reading. In the presence of the analyte, some different reading will be obtained; that reading then indicates the presence of the analyte and gives some information concerning how much analyte is present.

The difficulty that enters, and that makes this process of statistical interest, is the fact that any reading, whether the analyte is present or absent, is subject to noise and/or other sources of error.

Let us consider the situation in the absence of analyte, which is then exactly as pictured in curve A of Figure 14.2. In the absence of analyte, we know *a priori* that $\mu = 0$. The readings obtained, however, will vary around zero due to the noise. Indeed, when μ actually does equal zero, the presence of random noise will

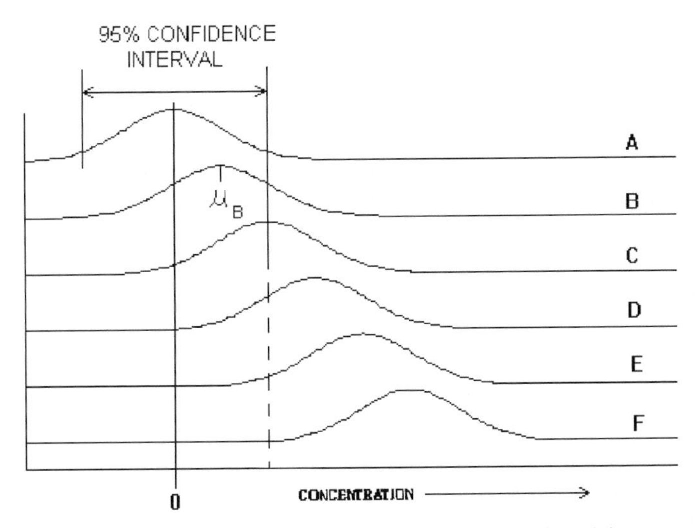

FIGURE. 14.2. As the true concentration increases, the probability of obtaining a measurement above the upper critical value for the null hypothesis μ also increases until, as shown in curve F, a sufficiently high measurement will always be obtained.

give rise to measured values (let us say, of an electrical signal that is otherwise proportional to the amount of analyte) that could even be negative. Ordinarily we would censor these; we know that negative values of a concentration are physically impossible.

For the sake of our discussion, let us not perform this censoring, but consider these negative values to be as real as the positive values. In this case, we might describe the situation as in Figure 14.1-A, and, to define a detection limit, decide to perform a hypothesis test, where the null hypothesis is $\mu = 0$ against the alternative hypothesis $\mu \neq 0$. In this case, failure of the null hypothesis would indicate that the analyte is present.

Clearly, this is not correct: even when the null hypothesis is true, we base our hypothesis test upon the fact that we will not find a value beyond the confidence limits, either above or below. In the presence of analyte, the distributions of the readings will be as shown by curves B–F of Figure 14.2.

Consider curve B, where $\mu > 0$ (but not by much). If, in fact, the concentration can be expressed as $\mu = \mu_B$, then the distributions of measured values will be as shown in curve B. If the null hypothesis $\mu = 0$ is used when the actual distribution follows curve B, then the probability of obtaining a value above the upper confidence limit will be approximately 25%, within the confidence interval 75%, and below the lower confidence limit effectively zero. Note that, even though the concentration is non-zero, we will fail to reject the null hypothesis more often than we will reject it in this case. As the true concentration (i.e., μ) increases, the situation is that it becomes more and more likely that the measured value will, in fact, be sufficiently high to allow us to reject the null hypothesis $\mu = 0$ until the point is reached, as shown in curve F of Figure 14.2, where the true concentration is so high that measured values will *always* cause rejection the null hypothesis, and we will always conclude that the analyte is in fact present. Inspect Figure 14.2, and take notice of the fraction of the area of each curve that lies above the upper confidence limit for the concentration $(\mu) = 0$ case; this should be convincing.

Thus, when $\mu > 0$, we will certainly never obtain a reading below the lower confidence limit, only values above $0 + a\sigma$ are possible. This asymmetry in the behavior of the system is exactly the condition that gives rise to the change in the true confidence level from 0.95 to 0.975 described earlier. Thus, even when we do not censor the values on physical grounds (which, by the way, would lead us to the same conclusion), the mathematics of the situation show that, in order to do a proper statistical hypothesis test at a true 95% confidence level, it is necessary to specify only the one confidence limit and, instead of performing the hypothesis test μ against the alternative hypothesis $\mu \neq 0$, the alternative hypothesis must be formulated: $\mu > 0$, and a one-tailed test used.

In order to implement this concept, it is necessary to know σ, or at least to have estimated σ by performing many analyses of a "blank", samples of a specimen (or set of specimens) that do not contain any of the analyte. Then, the upper confidence limit can be specified as an appropriate number of standard deviations of the distribution, which could either be Normal (if σ is known), or, more commonly, the t distribution (if σ must be estimated by the sample standard deviation S from the data).

We could then state that, when μ does in fact equal zero, the distribution is known, and furthermore, the distribution of the average of any number of readings is also known. We could then determine the criterion for stating that a single reading, or the average of a given number of readings, above the appropriate cutoff value, is an indication that the concentration of the analyte is non-zero.

Now, this is all fine and well when we can sit here and discuss distributions, and show pretty pictures of what they look like, and calmly talk about the probabilities involved. When you actually have to make a decision about a measurement, though, you do not have all these pretty pictures available, all you have is the measurement. (Well, that's not *quite* true, you can always look up this chapter and reread it, and look at all the pictures again.)

However, there are two important points to be made here: one obvious and one that is subtle. Curve A in Figure 14.2 shows the distribution and upper confidence limit of readings when $\mu = 0$. Curve C shows the distribution when μ is equal to the upper confidence limit of curve A. There is an important relation between these two distributions: the lower confidence limit of curve C equals zero. That is the obvious point. The subtle point is that there is an inverse relationship between μ and X (the value of a single reading): if a reading (rather than the true μ) happens to fall at μ_C, the distribution shown as curve C is then the distribution of probabilities for finding μ at any given distance away from the reading. Using this interpretation, if a reading falls at μ_C or above, then the chance of the true value being equal to zero becomes too unlikely, and so in this interpretation we also conclude that a reading higher than μ_C indicates a non-zero value for the true concentration.

The problem of defining a detection limit is that knowing how and when to do a hypothesis test, even using a one-tailed test, is only half the battle. This procedure allows you to specify a minimum value of the *measurement* that allows you to be sure that the analyte is present; but it does not tell you what the minimum value of the analyte concentration must be. As we noted earlier, if the concentration of the analyte is μ_B, there is a probability of well over 50% that the method will fail to detect it.

Suppose the true concentration (μ) is equal to the upper critical limit (for the null hypothesis $\mu = 0$) and which, by our sneakiness, is shown as μ_C. This condition is shown in curve C of Figure 14.2. In this case, half of the distribution lies above and half below the cutoff value, and the probability of detecting the presence of the analyte is still only 50%.

What is the minimum concentration needed? The answer depends upon the variation (i.e., the standard deviation) of the measurements *in the presence of the analyte*. This variation is not necessarily the same as the standard deviation of the measurements when no analyte is present. The curves of Figure 14.2 were drawn with the implicit assumption that the standard deviation of the measurements is independent of the amount of analyte present. In fact, many analytical methods are known where the error does depend upon the concentration (exercise for the reader: think of one from your own experience). In such a case it is necessary to determine the standard deviation of measurements in the presence of small amounts of analyte, as well as the standard deviation of measurements in the absence of any analyte.

Figure 14.3-A illustrates the situation in this case. When the true concentration, μ, is zero, the readings obtained will follow a distribution around zero, just as

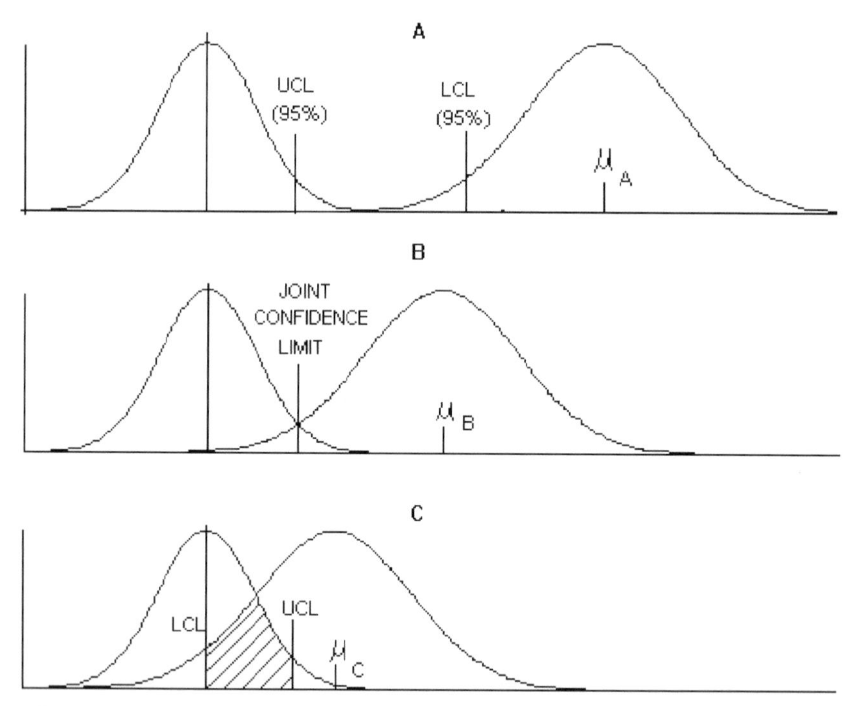

FIGURE. 14.3. (A) At very high concentrations, there is no overlap between the distributions of possible measurements, and the presence of analyte is readily distinguished from the absence of the analyte. Readings will essentially never occur between the two confidence limits shown. (B) When the concentration decreases to the point where the two confidence limits merge into a single value, it is still possible to distinguish unambiguously between the presence and absence of the analyte. (C) At still lower concentrations, there is a region where measurement can occur that will be consistent with both the presence and absence of the analyte.

before. This distribution will have an upper confidence limit beyond which any measurement indicates a non-zero concentration.

Clearly, in order to be sure of detecting the analyte when it is present, its concentration must be "sufficiently high". "Sufficiently high" in this case means that we are sure to obtain a reading that is high enough that there is a negligible probability of its falling below the confidence limit of the distribution around zero. In order for this to happen, the lower confidence limit of the distribution of measured values around the actual concentration must be equal to or greater than the upper confidence limit for the measurements around a μ of zero. This condition is illustrated by the distribution of measured values when the true concentration is μ_A. There is a region, between the confidence limits of the two distributions, where no measured values will occur.

In Figure 14.3-B the concentration has decreased to the point where the confidence limits of the two curves have merged into a single value. In the situation shown in Figure 14.3-B we could perform two hypothesis tests: $\mu = 0$, and $\mu = \mu_B$. Since the same actual value represents the confidence limit for both tests, this value separates the two possibilities. A continuum of measured values is possible in this case, but there is still an unambiguous demarcation between the presence and absence of the analyte.

If the actual concentration is lower yet, as shown in Figure 14.3-C, then there is a region between the two confidence limits where there is uncertainty, because a reading in this region would not cause rejection of either null hypothesis. When the concentration is still higher, as in Figure 14.3-A, there is no problem in distinguishing the two cases. Therefore, the minimum true concentration that will prevent an error in either direction is μ_B shown in Figure 14.3-B, where the true concentration equals the sum of the distances of the two confidence limits from their respective population mean. It is only when the true concentration is this high or higher that we can be sure that we will not erroneously claim that the concentration is zero because we accepted the null hypothesis: $\mu = 0$.

Thus, we see that there are at least these two possible definitions for detection limit: one specifies the minimum measurement that insures a non-zero concentration, and the other specifies the minimum concentration that ensures a non-zero measurement. In fact, there are an infinite number of possible values between these two that could be claimed. In each case, there is a different probability of obtaining incorrect conclusion from the two possible hypothesis tests. These intermediate possibilities are discussed at length in the references.

References

14-1. Clayton, C.A., Hines, J.W., and Elkins, P.D., *Analytical Chemistry*, 59: 20, (1987), 2506–2514.

14-2. Bergmann, G., Oepen, B.v., and Zinn, P., *Analytical Chemistry*, 59: 20, (1987), 2522–2526.

15

PHILOSOPHICAL INTERLUDE

Amidst all the discussion of distributions, significance tests, populations, confidence intervals, analysis of variance, and this-and-that-statistic, it is appropriate at this point, as it is perhaps appropriate to do every now and then, to look back and remind ourselves where all this comes from, and why it is worthwhile to spend so much time and effort (to say nothing of the forests that were hewn to provide the paper upon which we write this book) in discussing all these things.

As children, most of us were probably enticed into choosing chemistry as career by the possession of our first chemistry set, which allowed us to make colors and smells and "disappearing ink" and all sorts of good things. Now, as grown-up chemists, and more particularly as spectroscopists, we find that what we make mostly are measurements. The purpose of a measurement is to tell us something about the material on which we are making the measurement. We could make measurements of a physical property, such as density, hardness, or size. More commonly, we are interested in making measurements of chemical properties. The type of chemical property of interest depends upon the branch of chemistry: a physical chemist might be interested in measuring chemical bond length or perhaps heat of formation. An analytical chemist, on the other hand, is more likely to be interested in measuring the amount of some molecular or atomic species in a given type of sample (or perhaps in a given sample: the number of moon rocks brought back to earth does not quite qualify as a population).

As spectroscopists, we have learned to utilize, in one way or another, essentially the entire electromagnetic spectrum from DC to gamma rays, in the attempt to determine the composition of some part of the world around us.

However, no matter which part of the spectrum we use, or what kind of hardware we beg, borrow, steal, make or buy in order to achieve this goal, what we always wind up with is a number. The problem with a number is that alone it tells us nothing about the relation between its value and whatever property of our sample we are interested in.

Worse, experience tells us that when we make the same measurement more than once we do not even wind up with the same number. Whether we call this phenomenon "noise", or "error", or something else, we have the problem of deciding which of the several numbers available represents the quantity of interest.

Suppose we only made one measurement rather than several: are we better off or worse? In this case we avoid having to make a decision; the number at hand is the only one available that has a shot at telling us anything about what we really want to know. The problem now is that we have no idea whether or not the number does, in fact, bear any relation to the quantity of interest.

Mathematics to the rescue.

As we have demonstrated in previous chapters, numbers have properties of their own; furthermore, the more you know about the properties of numbers in general, the more you can say about the particular number that is the result of your measurement. We have been discussing the fact that certain kinds of numbers follow certain distributions. If you check books on statistics, you find discussions of all sorts of distributions: Poisson, binomial and hypergeometric distributions are commonly mentioned, as well as the ones we have discussed in the past several chapters (and some that we will discuss in the future chapters). While there is some overlap between the categories that we are about to create, there is a distinction that can be made between the above-mentioned distributions (Poisson, etc.) and some of the others [Normal, χ^2 (chi-squared), t, F].

We have been concentrating our discussions on the Normal and t distributions, and ignoring some of the others. The reason for this is the fact that, while the distributions from the first group (Poisson, etc.) describe the behavior of measurements under certain types of conditions, the distributions in the second set describe the behavior of *statistics*.

There is an important point to be noted here: just as a second measurement of a quantity will result in a different value for the quantity, a second calculation of a statistic will result in a different value of the statistic. No, that does not mean that $1 + 1$ will equal 2 one time and something else another time. What it means is that if you take some data (let us say 10 repeat readings of almost anything) and then calculate some statistics (let us say the mean and standard deviation) you will compute some value. If you then repeat the whole experiment, and take

10 more readings and calculate the mean and standard deviation from those data, you will get different values. Philosophically, you are almost no better off than you were with the data themselves.

Furthermore, the variation occurs whether the computations involved in generating the statistic are as simple as an arithmetic mean, or as complicated as the latest super-whiz-bang multivariate chemometric methods. The main difference between the simpler (generally univariate) statistics and the ones being promoted for chemometric usage is that the behavior of the simpler statistics have been investigated, and their distributions and properties known; while the more complicated ones (e.g., principal components, partial least squares, etc.) are both new and difficult to analyze with respect to spectroscopic data, so that their behaviors and distributions are not yet well characterized. We can only hope that the proponents of those methods are working on the problem.

In practice, however, there is an advantage to computing statistics. For the simple statistics, that have been thoroughly analyzed, the distributions of the statistics are known, and, more importantly, the properties of the statistics have been shown to be related to, and computable from, the properties of the data.

The other important fact about the usage of statistics is that among their known properties is the fact that they are better estimators of the "true" value of the quantity of interest than the raw data are. This has been shown rigorously for the statistics in common use, and, by analogy, we expect it to be true for the newer, more complex ones also. Only time and rigorous mathematics will tell how good this analogy is.

The problem with having only a single number available is that there are a good many hidden assumptions built into its use as an indicator of the properties of the material under study. The first assumption is that, in fact, a true value for the property of interest exists. Aside from the very deep and profound philosophical implications of a statement like this, which we will avoid like the plague, there are important cases where, in a practical sense, this assumption may not hold: heterogeneous samples, for example.

Even allowing the assumption that a "true" value exists, there is the question of whether the measurement we make is related to the quantity of interest. For example, it is well established both theoretically and experimentally that for transmittance measurements in clear solutions, where Beer's law holds, the Absorbance (defined as $\log(1/T)$) of a given species is proportional to its concentration. In reflectance spectroscopy, there is no rigorously derived relationship corresponding to Beer's law; in practice it is found that a quantity corresponding to and also called absorbance (defined as $\log(1/R)$) is also proportional to the concentration of the various species within a material. However, this relationship is not as clear as in the case of transmission measurements, and many other expressions are sometimes used instead of log(1/R), the Kubelka–Munk formula, for example.

Going one step further yet, often the property of interest is itself poorly defined, or, even if well defined, does not necessarily bear a clear relationship to the chemistry or spectroscopy involved. In practice, many of these cases work well despite lack of knowledge of the fundamental science involved, but they are based on an empirical foundation. It is not possible to tell *a priori* whether a given measurement of this nature can be made, one can only try it and see.

Finally, even allowing that all the prior assumptions are valid, there is the question of the noise, or error, inherent in the measurement itself, which we need not belabor here: for every instrumental or analytical technique there are usually long lists of possible sources of error that the analyst must guard against, often the errors depend upon the nature of the sample to be measured.

Consequently, no one would take a brand-new instrument out of its box, plug it in, present an unknown sample, and expect the number produced to have any meaning; such a procedure flies in the face of common sense. Even assuming that the instrument is not damaged or misadjusted there is no way to tell how much of the number is due to truth, and how much due to error or departure from truth.

At the very minimum, making the measurement twice will allow the analyst to determine that at least some of the errors are small, so that "truth" cannot be too far away from where the number says it is; but from one number alone even this determination is impossible.

In practice, of course, we prefer many numbers. Often, these numbers are collected in order to produce what is known as a "calibration"; other times the numbers verify one or more aspect of correct instrument operation. In any of these cases, the underlying purpose of collecting the numbers is to determine the distribution of results that are obtained under the specified conditions, whether or not that purpose is recognized. The recognition of this point brings with it, however, the potential to formalize the procedures; then one can not only calculate confidence limits for the various quantities measured (including the final answer), but can also apply one of the various "statistically designed experiments" techniques to allow collection of the data in an efficient manner. Use of those techniques simultaneously provides two other benefits also: they minimize the error portion of the number, and they allow an estimate to be made of how large the remaining error is.

Once having determined the necessary distributions, further measurements have meaning, even when they are only single numbers themselves. This is because, having determined how far away a number can be from truth, we have also determined how far away truth is from the number that we have in hand, that is what we could not do from the single number alone.

Since any measurement can be seen as being equal to the "true" value of the quantity of interest plus error, making a single measurement does not change the error, but it limits our ability to determine what the error is. There seem to be some misconceptions about this: sometimes an opinion is expressed that if only one measurement was taken, it must have no error, since the error cannot be calculated. As we showed in Chapter 13, this is clearly incorrect; if the error cannot be calculated or is indeterminate, it means only that the error is unknown, not that it is non-existent.

16

Biased and Unbiased Estimators

Moving right along, we expect that by now our readers understand and accept the fact that the key to understanding the science of statistics and what it says about the behavior of anything that contains a source of random variability is the fact that nothing is constant. Most chemists are at least exposed to this concept, but perhaps in a very limited way. If our own experience is typical, then what is taught in school is simply the fact that data can contain variability (usually called "error"), and the variability of the data can be quantified. Back in the days before computers or even pocket calculators were readily available, this quantification was performed on the basis of the "mean deviation" or the "relative mean deviation". Maybe something called "standard deviation" was mentioned, but at an elementary level it was considered "too complicated and difficult to calculate" by the manual methods then available.

This view is explainable in part by the limitations of the technology available at that time, but there is more. Part of the rationale for this view is more fundamental, and still hangs on today, even with (or perhaps because of) modern computational facilities. Spectroscopy, and science in general was, and still is, considered a deterministic discipline. Beer's law and other scientific laws are based upon *deterministic* considerations of the effects of various parameters. With the advent of widespread computerization of instruments, and the ready availability of computer programs for implementing complicated and sophisticated multivariate methods of data handling, we are entering into an era where results are sometimes explainable only by *probabilistic* considerations. The study of such phenomena has long been the province of mathematicians and statisticians, who have laid a strong groundwork for understanding this type of behavior in the twin disciplines of probability theory and statistics. Engineers have found the study of these disciplines very fruitful, for example, in explaining the effects of noise on data transmission and weak-signal detection. Perhaps, it is now time for a fundamental change; time for spectroscopists, and chemists in general, to start to make more widespread use of these other fields of knowledge; time, maybe, to learn about and apply statistics to current problems on a routine basis.

Measures of Variability

Nowadays, of course, the ready availability of computers and pocket calculators makes the computation of standard deviation easier and more straightforward than the computation of "mean deviation" (or any of the other measures of variability) used to be. We might also note parenthetically that this use of calculators also diminishes the chance of making a mistake in the arithmetic, or of reading a number incorrectly (data entry errors still can occur, however).

The student is left with the impression that the computed measure of variability, whether it is the standard deviation or one of the other measures of variability, is an accurate and reliable indication of the amount by which the data can vary, right? Wrong (of course, or we would not have anything to write about now).

In Chapter 15 we went into quite a diatribe concerning the fact that when statistics are calculated from data that have a random or variable component, the statistics themselves will also have a random component. This is true of the standard deviation as well as all other statistics. In fact, the standard deviation is by far the most variable statistic that we have yet considered. There are others that are even worse, but we have not gotten to them yet.

As we have mentioned many times, the variability of a random variable must be discussed in terms of its distribution. What, then, is the distribution of standard deviations?

Well, we have not made a thorough literature search to find out, but strangely enough, this point is never discussed in the commonly available treatments. Standard deviation is probably distributed as χ, but we are not sure, because mentions of the χ distribution are as rare as those of the distribution of standard deviation.

What can be found, in overwhelming abundance, are discussions of the χ^2 distribution, which describes the distributions of variance, the square of standard deviation.

Why is this? Why do statisticians place so much emphasis on variance, a rather abstruse measure of variability, rather than on standard deviation, a much more understandable one and one that has the enormous practical advantage of having the same units as the quantity being measured? Well, there are a couple of reasons.

The first reason is a practical one. We showed in Chapter 9 that variances are additive, and subsequently discussed some of the consequences. While it is reasonable in certain circumstances to compare standard deviations by dividing one by another, if you add standard deviations together you do not get a number that can be assigned meaning or that says anything rational about the data.

Variances, on the other hand, are much more flexible. In any circumstances where standard deviations can be multiplied/divided, so can variances. In addition, the sum or difference of variances has a definite meaning. If two or more sources of variability act on a measurement (or other datum), then the total variance of the measurement is equal to the sum of the variances due to each of the individual sources. Thus, use of the variance has great practical advantage.

The second reason is a theoretical one, but most statisticians agree that it is probably the more important one. In a philosophical sense we must agree with them, because having a sound theoretical basis for a science is almost a prerequisite to making progress in that field.

This reason is something we have mentioned before: the fact that the sample variance (S^2) is an unbiased estimator of the population variance (σ^2), while the sample standard deviation (S) is a biased estimator of the population standard deviation (σ). Let us correct a misconception that we may have left in our earlier discussion. At that time we were discussing the difference between using n and $n - 1$ in the denominator term of the standard deviation calculation, and the appropriate conditions for using each one. We somewhat loosely stated that using n in the denominator resulted in a biased estimator for σ while using $n - 1$ produced an unbiased estimator. We did at that time point out that the considerations were really correct for variances rather than standard deviations, but that was in a parenthetical note and may have been missed. At the time we were concerned with other, more basic, concepts and did not want to belabor the point or confuse the reader, especially since the difference between using n and $n - 1$ is much greater than the amount of bias in the standard deviation. Now, however, it behooves us to consider the situation in more detail.

Biased Versus Unbiased

This brings up two questions: first, what does it mean to say that S is a biased estimator of σ; and second, does it make sense?

Let us address the first question first (sounds good). We originally brought up the question of biased versus unbiased estimators in Chapter 3, and rather loosely defined an unbiased statistic as one that tends to cluster around the corresponding population parameter. Now the time has come to reconsider the question and provide a more mathematically rigorous definition.

We will not prove these definitions. Rather, for the proofs we refer the interested reader to the book by Hald (16-1), and our discussion will follow his book. This book, by the way, is a most excellent source for studying the math supporting the field of applied statistics. The proofs are rigorous without using overwhelmingly heavy mathematics, the explanatory discussions are almost as good as ours, and the book is very comprehensive indeed. The main limitation is that while explanatory examples are used, they are generally taken from fields other than spectroscopy; indeed, few chemical examples are used at all. Also, do not be put off by the way Hald uses symbols. Using ξ instead of μ to stand for population mean and a very fancy $\mathscr{M}$ to stand for "take the mean of" can be scary the first time you see it, but it is really not all that bad. This book is highly recommended.

In earlier chapters we used computer programs to compute statistics from all combinations of data from a defined population to demonstrate properties of whatever statistics we were discussing. The population we used was the integers 1–10. It was the clarity of Hald's mathematical proofs that provided the idea for this approach. On the other hand, because of the rigor of the mathematical proofs, they have a generality that is missing from our own demonstrations.

Thus, for example, on page 208, Hald defines an unbiased statistic as one for which the mean of all the values of the statistic, computed for all possible combinations of members of the original population, equals the corresponding population parameter. Of course, Hald presents this definition as a mathematical definition; we have translated it into words.

Hald's definition differs from our previous use of this concept in several important ways. First, when we used our computer program, we always computed statistics based on some fixed sample size, n, while the mathematical definition computes the mean of all possible combinations of samples of all sizes. Second, as a mathematical definition, it applies to all cases, including discontinuous as well as continuous distributions, and more importantly, to populations that are infinite in extent as well as those that are finite.

Thus, for example, on pages 199–203 we find the proofs that the mean of all sample means is the population mean, and that the mean of all sample variances (using $n - 1$ in the denominator) equals the population variance. The former proof is rather straightforward; indeed, we offer it as an "exercise for the reader". The latter proof is not nearly so simple. However, since the definition holds for both these statistics, they are both unbiased estimators.

Now, does it make sense that the standard deviation is a biased estimator when the variance is unbiased? It almost seems intuitively obvious that if the sample variance is an unbiased estimator of the population variance, then the sample standard deviation should also be an unbiased estimator of the population standard deviation. Ah, but no. Consider that the transformation from standard deviation to variance is non-linear; then you can begin to see the problem. In mathematics, one counter-example is sufficient to disprove a hypothesis; let us take this approach to understanding the question of bias in standard deviations.

Let us imagine a population whose standard deviation is 10; then the population variance is exactly 100. Let us further imagine that there are only two possible samples from this population, whose variances are 90 and 110. (Offhand, we cannot figure out how to construct the details of such a population; nevertheless, assuming it exists will serve for our example.) The mean of all sample variances equals the population variance, in accord with the proof that sample variance is an unbiased estimator.

The sample standard deviations from this population, however, are 9.486833... and 10.488088..., and the mean of those two numbers is 9.98746... Thus, the mean of the standard deviations of all samples from the population does not equal the standard deviation of the population (which is exactly 10), so this is a biased statistic.

How did this come about? The non-linearity of the conversion between standard deviation and variance caused the higher value to decrease more than the lower value, and also more than the population value did. This explanation is fairly general, and accounts not only for the existence of bias in standard deviation calculations, but also for the fact that the bias decreases as more and more data are included.

To understand this last point, we have to bring in two other concepts that we have discussed in previous chapters. The first is the central limit theorem, which, let us recall, states that the sum (and therefore the mean) of samples from any distribution approximates a Normal distribution more and more closely as more and more samples are included in the sum. This holds for the distribution of standard deviations as well as any other statistic; therefore, for large values of n, the distribution of standard deviations approximates a Normal distribution. The important point here is that, for large n, the distribution becomes symmetric.

The second important concept is that the standard deviation of the distribution of means becomes smaller and smaller as more values go into the computation of each mean. As the distribution gets tighter, differences in values become smaller and smaller, therefore the non-linearity has less range to act over. Therefore, while in principle the standard deviation is always biased, the bias becomes smaller and smaller as more and more values are included.

This discussion will help us to understand some of the concepts of error in spectroscopic data to be discussed in later chapters. For example, the peculiarities of artifacts found in spectral data can be explained by basic statistical principles. We will learn that a detailed understanding of biased and unbiased estimators, as well as a thorough explanation of analysis of variance, will allow us to separate and test the different sources of variation in spectral and other analytical data.

The major variations involved in analytical measurements include random errors in a measurement technique and actual variation in analyte concentration due to sampling variation. Each of these major variance sources can be further subdivided into its own set of variances. We will not be able to expand and discuss this list here and now, but ask that our readers bear in mind that the variations observed in analytical data are due to the summation of the effect of several types and sources of variation, which can then be explained and explored scientifically. This exercise then leads to a better understanding of the meaning of measurement.

References

16-1. Hald, A., *Statistical Theory with Engineering Applications*, Wiley, New York, 1952.

17

THE VARIANCE OF VARIANCE

In Chapter 16 we brought up the fact that the computed standard deviation, being itself a statistic, is also variable in the sense that when it is computed from one sample you will obtain one value, and when computed from a different sample, even though taken from the same population, you will obtain a different value.

The same is true, of course, for the variance, the square of the standard deviation. The obvious question that arises, then, is: how much does this statistic itself, the variance, change from one sample to the next? What is the variance of the variance? We say this in a humorous manner, as is our tendency, but of course it is in fact a serious and important question.

The Formula

Now Hald, as is *his* tendency, has calculated this for us and presented the answer in as straightforward and clear a manner as a complex subject can be described (17-1). The formula for the variance of the variances is (using Hald's somewhat non-standard notation):

$$\mathscr{V}(s^2) = \frac{\sigma^4}{n}\left[\frac{\mu_4}{\sigma^4} - \frac{n-3}{n-1}\right] \tag{17-1}$$

where $\mu_4 = \mathscr{M}\{(X - \xi)^4\}$.

Now, Equation 17-1 has a number of notable features. The one of most direct interest to our immediate question is the fact that the variance of the variance increases as the fourth power of σ. This is, indeed, a very strong dependency. We will shortly look at some of the numbers associated with this, to get a better feel for how much the variance does, in fact, vary.

Another notable feature is the fact that the variance of variance depends upon n, the number of data included in the computation of s^2 as well as on the variance itself.

A third item that is actually somewhat less important, but more interesting because it is unusual, is the expression that is included, that is represented by the symbol μ_4. This expression represents the property of data usually called the *kurtosis*. Kurtosis, along with skew, is often included in texts on data handling, statistics, and chemometrics—see, for example, the book about statistics by Box *et al.* (17-2), a highly respected text. Now, what is notable by its presence in most of these discussions is a long description of the calculation of these higher-order "moments" as they are called. Correspondingly, what is notable by its absence in most of these same discussions is any description of when, where, and how these higher-order moments can be used. So, the unusual character of the appearance of kurtosis in Hald's treatment is the somewhat refreshing fact that it demonstrates a *use* for a higher-order moment—a rare circumstance indeed!

A fourth notable characteristic of variance is the fact that the distribution of variance is asymmetric. In Chapter 5 we plotted the distribution of standard deviations obtained from a finite population (see Figure 5.2): the population of integers from 1 to 10. Notice how long the tail is on the high side of the distribution. Now imagine how long the tail would be if that distribution were stretched so that each bar of the histogram was moved to where it was placed on the X-axis at a point corresponding to the square of its current value. Even better,

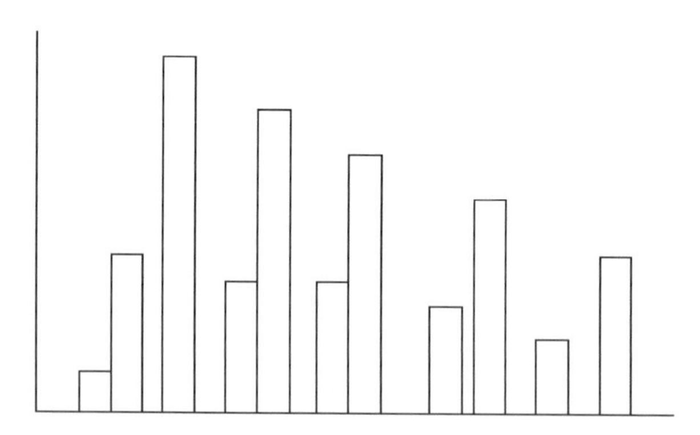

FIGURE 17.1. Histogram of the distribution of variances of samples from the population of integers 1–10 (data from Chapter 5).

you do not have to imagine it, we will show you: look at Figure 17.1, where we have done the stretching for you.

Now you have some idea of what the distribution of variances is like. Sure, there are differences between the particular distribution of variances that we presented, and the general one, which is valid for most real data. One difference, of course, is that the distribution of variances from our limited (finite) population is a set of discrete values that is represented by a finite number of discrete bars, while the general case is represented by a continuous distribution containing an infinity of possible values. Similarly, the standard deviations that are calculated from the values in our test population are limited to a well-defined and actually rather confined range. In the continuous case, the (theoretical) distribution actually extends to infinity.

The overall features of the "real" case are well described (not perfectly, perhaps, but still pretty well) by the limited case. There is a maximum around the "expected" value, indicating that the expected value is the one most likely to be found in any single instance in which this statistic is calculated. Lower values than the expected value can occur, but are limited by the fact that standard deviation (or the variance) cannot be less than zero. In the finite case, where the values are discrete, the variance actually was zero in some cases. In the continuous case, the probability of obtaining a value of zero itself falls to zero. However, the more important aspect of the situation, in either case, is not the actual limit, but rather the confidence limit, which is above zero in both cases. That takes care of the low end.

At the high end, the discrete finite case has a discrete limit, just as it does at the low end. The continuous case, however, has no upper limit. In theory, the distribution extends to infinity. The probability of obtaining a high value becomes vanishingly small, of course, but the curve itself goes on forever. At the high end, then, the only meaningful limiting value is the upper confidence limit, which extends upward proportionately along with the tail of the distribution curve.

The name of this distribution, which describes the behavior of variances, is the χ^2 (chi-squared) distribution, and it has unique and interesting properties of its own.

The first of these properties that we point out is the fact that there is not a single χ^2 distribution; there is a whole family of distributions, just as there is a family of t-distributions. The properties of each distribution depend upon the number of degrees of freedom in that distribution.

The second property is that the confidence limits for χ^2 depend upon the number of degrees of freedom in the data that the particular value of χ^2 is being calculated for. This is clear in the formula for χ^2 (which is complicated and we choose not to present it here) and this fact can also be seen in tables of χ^2. At this point it will be constructive for our readers to get some exercise by standing up and taking the book of statistical tables from the shelves (the one we recommended acquiring back in Chapter 6) so that you can follow along. We are going to refer to them several times in the rest of this chapter, as well as the rest of the book.

Note first of all, that the table is laid out so that one axis (usually the horizontal one) is labeled with probability levels, while the other (usually the vertical one) is listed in order of increasing number of degrees of freedom. Symbols used to designate degrees of freedom vary. Notations such as d.f., (lower case) f, or the Greek letter ν are often used. The symbol we will use for degrees of freedom is the lower case f.

Whichever symbol is used, the table entries in that column invariably start with 1 and continue with every integer until at least 20. At some point the table will increase the interval at which the entries are presented.

Now look at the entries themselves. Note that they increase both as you read down the table (larger numbers of degrees of freedom), and as you read across (higher probabilities). What is the importance of this?

As f increases, the lower confidence limits (LCLs) increase rapidly. Not only is the expected value increasing, but also in addition the lower confidence limit has to run to catch up to it, since it must be closer to the expected value at higher values of f. The upper confidence limit (UCL) is subjected to two opposing factors: the expected value is increasing, but the upper confidence limit has to approach the expected value. The net result is that the UCL also increases, but not as rapidly as the LCL.

So how does this relate to using the χ^2 tables? Since the 50% point equals f as well as representing the expected value of χ^2, each row of the table can be normalized by dividing each entry in that row by f. This would make the expected value of the statistic always be unity, and every other entry would then be the multiplier for s^2 that defines the value of s^2 corresponding to that probability level. In fact, tables are available where this division has already been done for you, and the tabled values are χ^2/f. Such tables are otherwise identical to the χ^2 tables themselves.

A visual illustration of the variability of the standard deviation, and the effect of the number of data points that goes into calculating the standard deviation, is presented in Figure 17.2. This figure was generated by rescanning a 100% line 100 times, then computing the standard deviation at each wavelength, for the full set of 100 scans, and then for different size subsets of these spectra. This data has a population standard deviation (i.e., σ) of approximately 0.00015. Figure 17.2-A shows one of these typical spectra; Parts B, C, and D show the calculated standard deviation(s) for the set of data, taken over 10, 20, and 100 readings, respectively. In accordance with Equation 17-1, all three sets of standard deviations cluster around the population value, but the clustering gets tighter and tighter as more and more readings are included in the calculation,

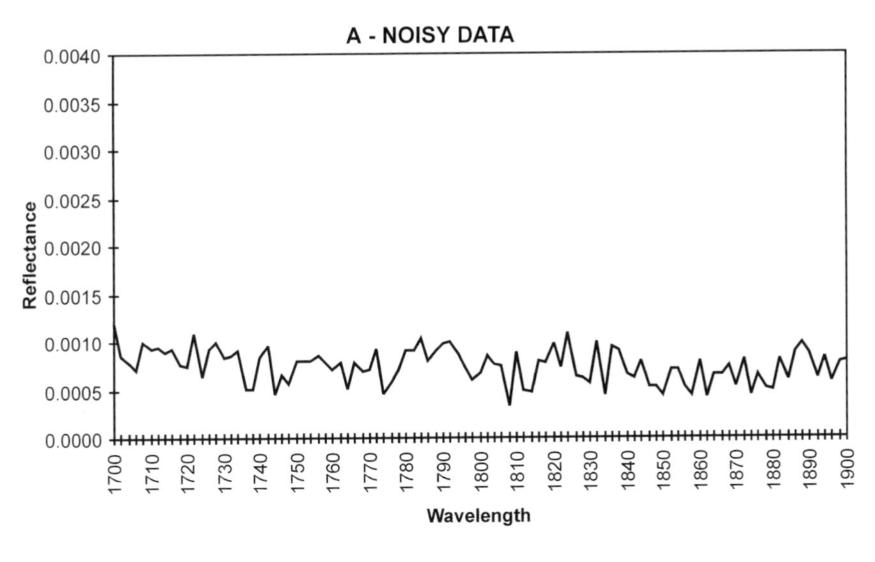

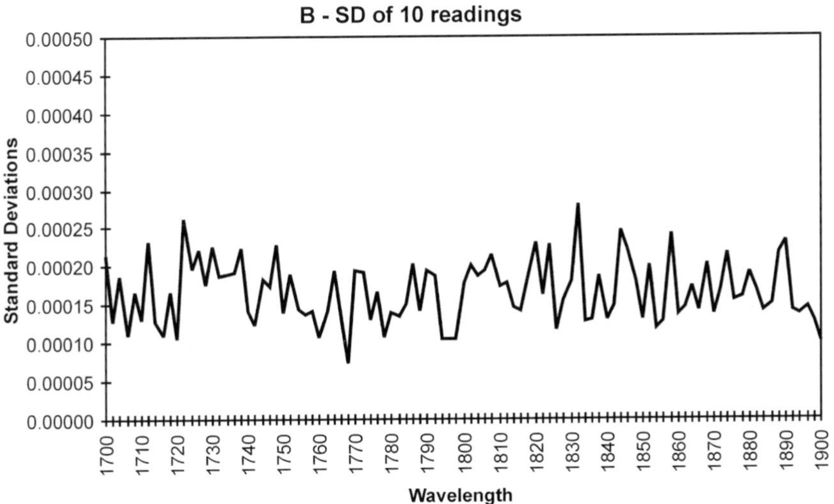

FIGURE 17.2. The effect of the number of samples used in the calculation of standard deviation. (A) One of a set of spectra with a standard deviation of approximately 0.00015. (B–D) The calculated values of standard deviation when 10, 20, and 100 readings, respectively, are included in the standard deviation calculation. While all the values tend to stay around the population value, the clustering gets tighter as more readings are included in the calculation.

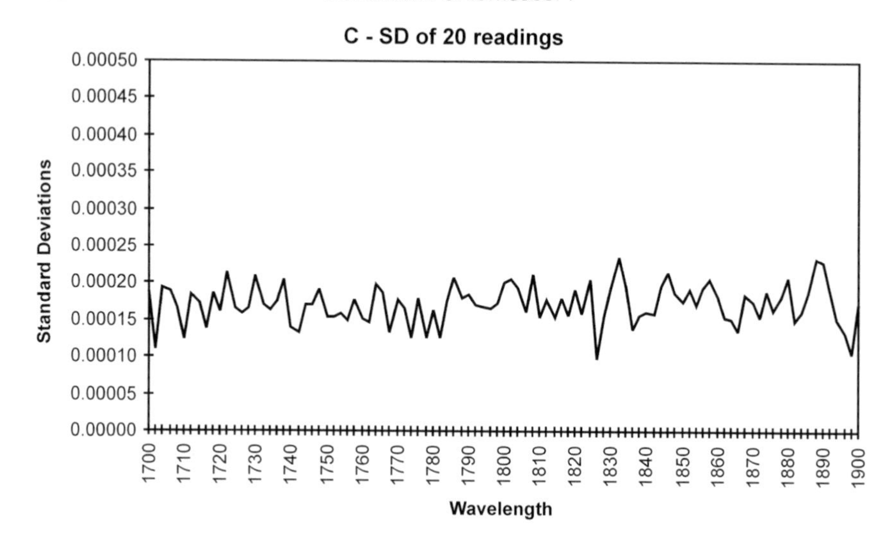

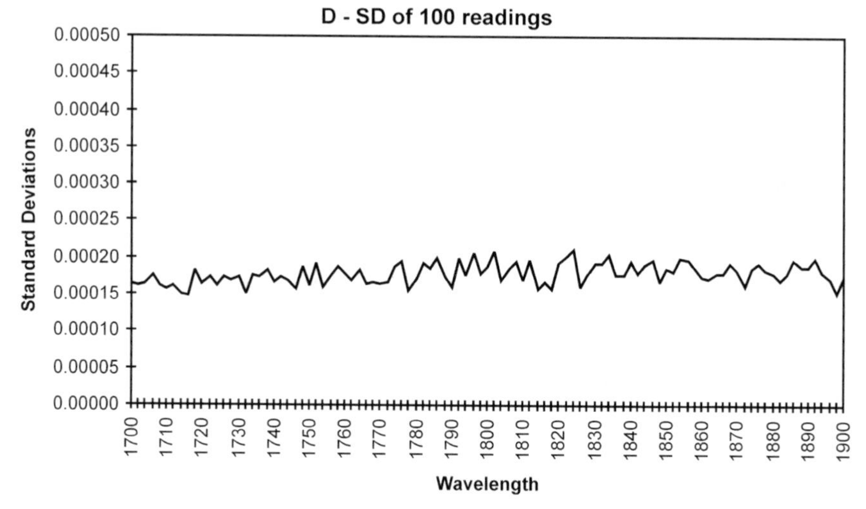

FIGURE 17.2. (*continued*)

and that, of course, is the point of this whole exercise. By the way, also note that the mean value of the data (approximately 0.001) has nothing to do with the values of standard deviation calculated—we would have achieved the same result if the mean were zero, or 100, for that matter. Whether you use tabled values of χ^2/f, or compute this value for yourself, the point is that the resulting values are the multipliers that designate the confidence limits of the variance, and taking the square root of those values gives the corresponding confidence limits of the standard deviation.

TABLE 17-1
Ratios of upper confidence limit of standard deviation to lower confidence limit.

Degrees of freedom	95%	99%
1	70.88	
2	12.02	32.55
3	6.57	13.35
4	4.79	8.47
5	3.92	6.37
10	2.51	3.41
20	1.88	2.31
30	1.67	1.97
40	1.55	1.79
50	1.48	1.68
100	1.32	1.44

What we want to point out in this regard is the amount of variability that the standard deviation is subject to. For this purpose, we have created Table 17-1, where we list a characteristic of the standard deviation not often presented in this manner. The entries in Table 17-1 are the ratios between the UCL and the LCL for standard deviation, at the 95 and 99% confidence level of the χ^2 distribution, for several degrees of freedom.

Will you look at that! They say that if you have a lot of samples, you get a good estimate of standard deviation. But look at what Table 17-1 tells us! With 10 samples, the standard deviation can vary by a factor of $2\frac{1}{2}-1$, purely by random chance alone! Well, we guess that means that 10 is not enough samples. But even 31 samples (for 30 degrees of freedom) give a ratio of standard deviation of 2–1 (at the 99% confidence level). And for the die-hards, look at the entry for 100: you can still get a variability of almost $1\frac{1}{2}-1$.

Now *that is* a variable statistic.

Understanding the χ^2 distribution will then allow us to calculate the expected values of random variable which are normally and independently distributed. An important note here is that in least-squares multiple regression, or in calibration work in general, there is a basic assumption that the error in the response variable is random and normally distributed, with a variance that follows a χ^2 distribution. Such factors are important to remember when utilizing calibration techniques. We will discuss this, and other related points when we get up to discussing calibrations; we include these notes here as a forewarning of factor that should not be ignored.

References

17-1. Hald, A., *Statistical Theory with Engineering Applications*, Wiley, New York, 1952.
17-2. Box, G., Hunter, W., and Hunter, J., *Statistical Theory with Engineering Applications*, Wiley, New York, 1978.

18

HYPOTHESIS TESTING OF CHI-SQUARE

In Chapter 17 we reviewed just how variable the standard deviation is, when considered as a statistic that has different values when computed from different samples of a population of random errors. What we found was that the standard deviation was quite variable indeed, since the variance, from which the standard deviation is computed, is proportional to χ^2. Strictly speaking, the quantity that follows the χ^2 distribution is $(n-1)s^2/\sigma^2$, a formula we will have need for when we actually do hypothesis testing using variances.

Since this is the case, even though the standard deviation of standard deviations can be computed, it is not permissible to use the common approximations, such as the one stating that ± 2.5 standard deviations includes 95% of all standard deviations. The reason for this is that the approximation is valid only for Normal distributions, and standard deviations do not follow that. The only proper way to determine confidence limits, and do hypothesis testing on standard deviations, or on variances, is to use the proper table, i.e., a table of percentage points of χ^2. If variances are of interest, the table is used directly; if standard deviations are of interest, then the square roots should be taken *after* the confidence limits for variances are determined.

There is nothing intrinsically wrong with putting the results in terms of standard deviations, as we have just described, especially when it aids understanding by putting the answer in terms of familiar units. However, there are a good number of statisticians who would almost have apoplexy at the thought of doing such unorthodox conversions, and there is one good justification for their concern. Statistics, as we have mentioned in the past, is fraught with pitfalls even when following the straight and narrow path. When unorthodox manipulations are performed, it is imperative to be extra cautious against performing operations that are not permitted for standard deviations, even though they are permitted for variances. Clearly, it may be all too easy to get confused as to what is what.

So, if standard deviations are so much trouble, and both variances and standard deviations are themselves as variable as we have found them to be, what good are they at all? Well, would you believe that despite all those awful characteristics we have just discussed, standard deviation is the best possible way to compute the variability of the data? It is a fact. The statisticians have shown that of all possible ways to measure variability of data subjected to random influences (other measures, for example, include the range and the mean deviation), the variance is the one that has all the properties that statisticians consider desirable. What properties are these? Well, they are the ones that we have brought up now and then: it is unbiased, it is a maximum likelihood estimator, and finally, believe it or not, and more to the point here, despite how variable it is, the variance is the measure of variability that itself has the least variance of all possible estimators of variation. We repeat: just as the variance of variance can be computed, so can the variance of range, mean deviation or any other measure of variability, and the variance of the variance is smaller than the variance of any of these other measures.

That is the second reason variance is so popular among statisticians (the first, you recall, is that variances add). Even though standard deviation and variance are themselves so variable, they are the measures of variation that provide the smallest possible confidence intervals against which to perform hypothesis tests. What can we use such confidence intervals for? Well, for example, they can be used to characterize the amount of variation that might be seen on the measurement of noise level of an instrument.

From this, one might expect that this is how instruments could be compared—in fact we can almost hear the heavy breathing out there at the prospect. However, χ^2 is *not*—repeat, *not*—the proper distribution for making such comparisons—see how easily you can fall into a statistical trap!

It is, however, the proper distribution to use to compare an instrument against its own specification. The "spec" value in such a case assumes the role of the population value for the noise level; then it becomes possible to perform hypothesis tests that compare actual measured values of noise against that population value. For example, let us suppose that the specification value for electronic noise at some critical point in an instrument is 2.5 units—we do not state any units because for our current purposes here that is not important; it could be voltage measured in millivolts, transmittance or reflectance measured in milli- or micro-absorbance, or even the precision of the final computed answer of a quantitative analysis. In any case, let us say the instrument spec. is 2.5.

The manufacturer, of course, wants to verify that a given instrument is in compliance with his specification. Therefore, he collects some data and calculates the standard deviation and variance, due to the noise, of this quantity. Let us consider some possible outcomes of such measurements.

While doing this, we will take one more step to formalize the process of performing hypothesis testing. The formalism of hypothesis testing states that a hypothesis test should be carried out in a set of discrete steps, as follows:

(1) State the experimental question.
(2) State the null hypothesis, and the alternative hypothesis.
(3) Decide the level of significance (α, which $= 1 -$ the confidence level)
(4) State the test statistic (i.e., the standardized form that can be used to reject the null hypothesis)
(5) Find the critical value(s) for the test statistic (from statistical tables, usually).
(6) Collect the data (ideally, this is the proper place in the sequence for this step; in actuality this step is of necessity often done out of order).
(7) Calculate the test statistic from the data.
(8) State the statistical conclusion.
(9) State the experimental conclusion.

Scenario #1: Let us apply the formalism to the problem at hand. In this scenario, the instrument is measured to have a standard deviation of 2.93, using 50 readings, giving 49 degrees of freedom. Now we will follow the formalism:

(1) State the experimental question: Is the instrument performing within its specification?
(2) State the null hypothesis, and the alternative hypothesis:

$$H_0: \sigma = 2.5$$
$$H_a: \sigma \neq 2.5$$

(3) Decide the level of significance: $\alpha = 0.05$.
(4) State the test statistic: $\chi^2 = (n-1)s^2/\sigma^2$.
(5) Find the critical value(s) for the test statistic:

$$\text{LCL} = \chi^2(49, 0.025) = 31.555$$

$$\text{UCL} = \chi^2(49, 0.975) = 70.222$$

(6) Collect the data: $S = 2.93$.
(7) Calculate the test statistic:

$$\chi^2 = 49 \times 2.93^2/2.5^2 = 67.30$$

(8) State the statistical conclusion: the measured variance is not beyond either confidence limit and therefore there is no evidence from the measured standard deviation that the population value is different than the specified standard deviation.
(9) State the experimental conclusion: the instrument is performing within the specification.

The use of this formalism may seem a bit pedantic, but often as not it saves grief. We have noted several times the existence of pitfalls in the use of statistics. The formalism helps avoid some pitfalls by exhibiting what is going on with crystal clarity. Often as not, minor and seemingly unimportant changes in the nature of the problem can cause the proper statistical treatment to differ from that of the initial problem. Let us look at some variations on the question we have brought up.

Scenario #2: For our first variation, the test procedure has changed; the measurement is now performed using 300 readings to compute the standard deviation instead of 50, a not unreasonable number if data are automatically collected by a computer. Suppose the same value as above is obtained. Now the hypothesis test looks like this:

(1) State the experimental question: Is the instrument performing within its specification?
(2) State the null hypothesis, and the alternative hypothesis:

$$H_0 : \sigma = 2.5$$

$$H_a : \sigma \neq 2.5$$

(3) Decide the level of significance: $\alpha = 0.05$.
(4) State the test statistic: $\chi^2 = (n-1)s^2/\alpha^2$.
(5) Find the critical value(s) for the test statistic:

$$\text{LCL} = \chi^2(299, 0.025) = 253.9$$

$$\text{UCL} = \chi^2(299, 0.975) = 349.8$$

(6) Collect the data: $S = 2.93$.

(7) Calculate the test statistic:

$$\chi^2 = 299 \times 2.93^2/2.5^2 = 410.70$$

(8) State the statistical conclusion: the measured variance is beyond the upper confidence limit and therefore we have evidence that the measured standard deviation is different than the specified standard deviation.
(9) State the experimental conclusion: the instrument is not performing within the specification.

Well, of course, we stacked the cards on this one; you probably guessed that the same result, measured from a larger number of data, would make a difference in the answer. However, now let us consider another variation. In this case we are still computing from 300 data points, but now the computed standard deviation is 2.06.

Scenario #3:

(1) State the experimental question: Is the instrument performing within its specification?
(2) State the null hypothesis, and the alternative hypothesis:

$$H_0 : \sigma = 2.5$$
$$H_a : \sigma \neq 2.5$$

(3) Decide the level of significance: $\alpha = 0.05$.
(4) State the test statistic: $\chi^2 = (n-1)s^2/\sigma^2$.
(5) Find the critical value(s) for the test statistic:

$$\text{LCL} = \chi^2(49, 0.025) = 253.9$$

$$\text{UCL} = \chi^2(49, 0.975) = 349.8$$

(6) Collect the data: $S = 2.06$.
(7) Calculate the test statistic:

$$\chi^2 = 299 \times 2.06^2/2.5^2 = 203.1$$

(8) State the statistical conclusion: the measured variance is less than the lower confidence limit and therefore we have evidence that the measured standard deviation is different than the specified standard deviation.
(9) State the experimental conclusion: the instrument is not performing within the specification!

Now how could *that* happen? The answer is that we led you into one of the pitfalls of statistics. We deliberately misstated the question, which resulted in performing the wrong hypothesis test. Since we ran out of space in the original column at this point, we will show you where we went wrong, and what to do about it, in the next chapter. We will also discuss other factors that need to be considered in order to make statistical calculations conform to the real world.

19

MORE HYPOTHESIS TESTING

In Chapter 18 we left off with scenario #3 just as we came to the remarkable conclusion that an instrument that has a lower noise level than the specification value does not meet that specification. The reason we achieved this unexpected result was because we misstated the question we were trying to answer. Of course we did this deliberately, in order to create the conditions that give us the excuse to discuss the point at greater length. Let us examine this. Following the statistical formalism, let us examine the first two points as we stated them:

(1) State the experimental question: Is the instrument performing within its specification?
(2) State the null hypothesis, anld the alternative hypothesis:

$$H_0 : \sigma = 2.5$$

$$H_a : \sigma \neq 2.5$$

The first point implies that we do indeed want to know if the instrument under test conforms to the population value of noise for that instrument. What we found was that, for the noise level measured under the conditions specified, the noise was too *low* to have come from the stated population. Now, as a statistical hypothesis test that is a proper and valid question to ask, and that information may well be of interest, for example, to the engineers who are looking for ways to improve the instrumentation. As a method of qualifying instruments for use, however, it has led us to reject the instrument because that unit did not conform to the population by virtue of being too good, clearly the improper action.

The Right Question

The question we really wanted the answer to is: "Is the noise level of this instrument low enough to conform to the specification?" Stated this way, it becomes clear that we used the wrong hypothesis test, the one we should have used is a one-tailed test, rather than the two-tailed test we have applied to the problem. Now the hypothesis test appears as follows.

Scenario #4:

(1) State the experimental question: Is the instrument noise level sufficiently low as to conform to the specification?
(2) State the null hypothesis, and the alternative hypothesis:

$$H_0: \sigma < 2.5$$

$$H_a: \sigma \geq 2.5$$

(3) Decide the level of significance: $\alpha = 0.05$.
(4) State the test statistic: $\chi^2 = (n-1)s^2/\sigma^2$.
(5) Find the critical value(s) for the test statistic:

$$\text{UCL (upper confidence limit)} = \chi^2(299, 0.95) = 341.4$$

(6) Collect the data: $S = 2.06$.
(7) Calculate the test statistic:

$$\chi^2 = 299 \times 2.06^2/2.5^2 = 203.1$$

(8) State the statistical conclusion: the measured variance is less than the critical value and therefore there is no evidence that the standard deviation is higher than the specified standard deviation.
(9) State the experimental conclusion: the instrument is performing within the specification.

There are a number of seemingly minor but actually important differences between this hypothesis test, and the one of scenario #3. We will now point out and discuss these differences.

The first difference, of course, is the statement of the question. As we discussed earlier, the nature of the desired information dictated that the hypothesis to be tested was whether there was any evidence that σ, the population value for standard deviation, was higher than the specified value, rather than whether σ was equal to the specified value. Note, however, that properly stating the question is a crucial point. There are cases, and we pointed out one such, where showing equality (or lack thereof) is the proper way to proceed. In such a case, scenario #3 is indeed correct.

The second difference lies in the second point of the formalism. The null and alternative hypotheses have changed from equal/not equal to less/greater, to reflect the change in the question to be answered.

The level of significance and the test statistic for the hypothesis test is unchanged; the level of significance does not depend upon the nature of the test, but on external factors. In the absence of other information, a 95 or 99% confidence interval is usually chosen. The test statistic to use of course depends upon the information to be extracted from the data and is chosen *a priori*.

The critical value used is one of the crucial differences between the two cases. First of all, note that there is only one critical value listed. In the previous scenarios, we had both a lower and an upper critical value to compare our experimental value with. In the one-tailed test, however, we specify only one critical value, and the acceptance or rejection of the null hypothesis depends only on comparison with this single criterion.

A Tale of Two Tails

Now this brings to mind a question (or should, if you have been paying attention): why do we specify the UCL rather than the LCL? When you understand the answer to this question you will be home free on understanding the whole concept of hypothesis testing.

To this end, let us first review what hypothesis testing is all about. Hypothesis tests based on the χ^2 distribution have the same purpose as hypothesis tests based on other distributions, such as the Normal or t distributions. That is, it is the attempt to determine whether a given measured statistic could have come from some hypothesized population. In the case at hand, the hypothesized population is one that has a standard deviation of 2.5. Remembering that any given set of data taken from this population will not exhibit exactly that value of standard deviation, we still need to be able to make the determination. In order to do this, we fall back on the known properties of random numbers, and rely on the fact that we know the limits within which a set of data from that population can fall. If our measured data result in a statistic that falls beyond the limits, then we have evidence that the data did *not* come from the specified population.

Now to relate this general discussion to the question of how to set up the hypothesis test, specifically whether to make the rejection region the region above the UCL, as we did, or the region below the LCL: Making the rejection region those values of χ^2 above the UCL, we are looking to reject those instruments whose measured standard deviations are so high that the units they were measured on could not have come from a population whose σ was at or below the specification value. This was exactly what was desired, and from one point of view, is simply a formalization of common sense. Keep in mind, however, that a measured standard deviation that is sufficiently low that we do not reject the instrument along with the null hypothesis does not *prove* that the instrument is satisfactory. It only means that we have *failed* to prove it unsatisfactory—note the double negative: failure of proof is not the same as proof of non-failure.

On the other hand, if the rejection region included those values below the LCL, then by rejecting the null hypothesis we would indeed have found those instruments whose standard deviations were so low that their long-term performance would surely have been better than the specification value. This would identify those instruments that were unusually good, but if we (as a manufacturer, say) would sell only those instruments that passed such a stringent test, we would find that (assuming a 95% confidence interval and that the instruments were in fact just meeting the spec) only 5% of our production was satisfactory!

There are two points, however, that argue in favor of using the LCL. One is a valid statistical concern, the other is a perception (or a misperception, perhaps) of the meaning of statistical results. The first point is the fact that, whenever a statistic is calculated, it is, in fact, the best estimator of the corresponding parameter. Therefore, if the value calculated for S is between the specification value and the UCL, then there is a likelihood (that increases with the calculated value of standard deviation) that the long-term performance of that unit is indeed higher than the specification, even though the data available is insufficient to prove that such is the case.

The perception referred to is the fact that purchasers of instruments, when they test the units, expect to find that noise test results such as we are discussing are (almost) always less than the specified value. If σ for the instrument were in fact just equal to the specification, then fully 50% of the time the measured value would be above that value and the instrument performance would be considered unsatisfactory in the "real world" by the users.

Thus, for both those reasons, it is in fact necessary to make the instruments good enough that σ is far enough below the specified value, so that when a unit is tested the measured value of S always is less than the specification.

How Low Should You Go?

How much below the specification does σ need to be? Well, that depends upon how it will be tested. Interestingly, in this case the more stringent the test, the less margin for error is needed. This can be seen by calculating a few values of χ^2/f (at some fixed confidence limit). For example, at 95% confidence, for 10 degrees of freedom, $\chi^2/f = 1.83$, while for 50 degrees of freedom, $\chi^2/f = 1.35$. Therefore, if the instrument is tested with 10 readings, then the population value of the noise must be reduced to $2.5/1.83^{0.5} = 1.85$, while if the test is done with 50 degrees of freedom the noise need be reduced only to 2.15 units (remember, we are discussing these noise values in arbitrary units) in order for measurements to be below the specification value 95% of the time. Of course, one must be sure that others testing the units will use the same test procedures, with the same number of degrees of freedom!

Now, a passing rate of 95% may not be considered good enough, in which case the corresponding calculations may be performed at a 99% confidence level or even higher. Exercise for the reader: find the necessary value for the population value of noise (i.e., redo these calculations) based on a 99% confidence interval, so that 99% of the instruments will pass when the specification is 2.5. For more practice, try it at different specification values, and tests with different numbers of degrees of freedom.

For still more practice, consider that scenarios #1 and #2 in Chapter 18 were also based on a two-tailed hypothesis test. As we have shown, that was not the correct way to perform the test. As an exercise, reconsider those scenarios and determine the results from the proper, one-tailed tests. Are there any differences? Explain why some results are different (if any) and some not (if any).

20

STATISTICAL INFERENCES

In Chapters 18 and 19, we posed some questions and demonstrated how to answer them using the statistical concept of hypothesis testing. We also presented the full-blown formalism for doing hypothesis tests in all its glory. We even managed to show how the formalism can help to pose the right question for the experiment at hand.

The formalism varies somewhat, however, with the hypothesis test being used; since there are a number of variations on the ways in which the concepts we have been expounding can be put into practice. The many books about elementary statistics usually cover these variations in more or less organized formats. We have extracted the following list from one of these books that have the topics organized fairly well (20-1). This book, indeed, is particularly useful for this purpose. It not only gives each variation explicitly, but also, in addition, for each one it lists the nine steps in the formalism of the hypothesis test with all the minor variations appropriate for that particular hypothesis test. Even more than that, since it is a textbook it contains worked-out problems in the body of each chapter and also unworked-out problems at the end of each chapter (with answers at the end of the book, for checking). Of course, since the book is about statistics in general, it does not contain any applications to spectroscopy (although some problems dealing with other branches of chemistry appear).

Statistical Cases

Thus, considering only those statistics that we have already discussed in this book, the Z (means, normal distribution, σ known), t (means, normal distribution,

σ unknown), and the χ^2 distribution (standard deviations), we present the following list of generalized applications, extracted from Dixon and Massey (20-1). Basically, these constitute the generalized experimental questions that the science of statistics can answer using only the concepts of the Normal, t, and chi-squared distributions:

(1) Population mean is equal to a specified number when σ is known (two-sided test).
(2) Population mean is equal to a specified number when σ is known (one-sided test).
(3) Population mean is equal to a specified number when σ is not known (two-sided test).
(4) Population mean is equal to a specified number when σ is not known (one-sided test).
(5) Population proportion is equal to a specified proportion. While proportions have special properties, the proportion of items having a certain characteristic can be used in place of the mean in a large number of cases. This is useful when the characteristic in question cannot be assigned a physically meaningful measurement, but one can only ascertain whether an object has the characteristic or not (see page 100 in Reference (20-1) for more information concerning this point). An example might be whether a molecule has an absorbance band at a particular wavelength, when the strength of the band is ignored. The classic statistical example is the classification of objects according to whether they are "defective", but in general, any dichotomous classification scheme can come under this category.
(6) Population variance equals a specified number (one-sided test).
(7) Means of two populations are equal assuming they have the same σ when σ is known (two-sided test).
(8) Means of two populations are equal when σ is known (one-sided test).
(9) Means of two populations are equal when σ is not known (two-sided test).
(10) Means of two populations are equal when σ is not known (one-sided test).
(11) Means of two populations are equal when the variances are not equal, but known.
(12) Means of two populations are equal when the variances are not equal and are unknown.
(13) Difference between two population means is equal to a specified number, variances equal and known.
(14) Difference between two population means is equal to a specified number, variances equal but unknown.
(15) Difference between two population means is equal to a specified number, variances unequal and known.

(16) Difference between two population means is equal to a specified number, variances unequal and unknown.

Note that from number 11 on, we forbore to separately list one- and two-sided tests, in order to reduce the proliferation of cases. However, from the first 10 items it should be clear that each of the subsequent ones can be further subdivided into the one-sided and two-sided cases.

New Concepts

There are some new concepts introduced in this list. Previously we have considered only the relationship between measured data and a single population. Here we have introduced some new types of hypotheses that can be formulated. Of particular interest is the fact that the list includes comparisons between data that come from two populations. Other differences have to do with the way the correct standard deviation is calculated, a feature tied in with the fact that the correct value of S or σ must be determined from considering the appropriate contribution from both populations involved.

In writing this book we are torn between two conflicting desires. On one hand is our desire to expound each point we bring up fully, using both mathematics and expository discussion to explain everything as completely as possible (and to do it all in words of one syllable or less). Simultaneously we want to cover all the important concepts of statistics, so that we present both breadth and depth to our readers. On the other hand, we must recognize that we have only a finite amount of time and space, and need to keep moving along (not the least reason for which is to keep from boring you). Consequently, while we would like to derive the equations necessary to cover all the cases in the above list and to discuss their ins-and-outs, we feel we have to short-circuit the process, in order that we can continue on to other, equally important topics in statistics. We therefore present all the information compressed into Table 20-1. We recommend most strongly that you refer to other, more comprehensive works to fill in the gaps that we know we are leaving. We hope that our previous discussions, combined with the information we present here will suffice to start you on your way. We do honestly feel that this is only a start. Even for the topics we have covered, we have only scratched the surface. There are many books available that discuss statistics in greater depth and/or breadth. Before we continue to present more topics, we do want to state that, while we are trying to give our readers the best introduction of which we are capable, we recognize that it is only a very brief overview. We will feel we have succeeded if you, our readers, have gained enough insight into statistical thinking to be able to pick up a regular statistics textbook and be able to follow it and pursue by yourself those subjects that you find interesting or useful.

TABLE 20-1
Summary of computations.

	Population variance known	Population variance unknown
Population mean equals specified number[1]	$Z = (\bar{X} - \mu)/\sigma$	$t = (\bar{X} - \mu)/S$
Means of two populations are equal, when variances are equal[2]	$Z = (\bar{X}_1 - \bar{X}_2)/\sigma_p$	$t = (\bar{X}_1 - \bar{X}_2)/S_p$
Means of two populations are equal, when variances are unequal[2]	$Z = (\bar{X}_1 - \bar{X}_2)/\sigma_p$	$t = (\bar{X}_1 - \bar{X}_2)/S_p$
Difference between two population means equals a given value[1,3] (Note 2 also applies when variances are unequal)	$Z = \bar{D}/\sigma_d$	$t = \bar{D}/S_d$
	Population mean known	Population mean unknown
Population variance equals a specified number[4]	$\chi^2 = s^2/\sigma^2$	$\chi^2 = s^2/\sigma^2$

1. σ and S both refer to the corresponding distribution of means, which is determined from the corresponding values for individual data by dividing by $n^{1/2}$.

2. σ_p and S_p are called the *pooled* standard deviations, and are obtained by combining the individual values of the two populations to obtain what is sometimes called the RMS value. Even when the population variances are equal, we must pool them to account for the possibility of computing $\bar{X}$ from different sample sizes. The formula for computing the pooled standard deviation (in the most general form) is:

$$\sigma_p = \sqrt{\frac{\sigma_1^2}{N_1} + \frac{\sigma_2^2}{N_2}}$$

The formula for computing the pooled sample standard deviation, S, is exactly analogous. Also, when $N_1 = N_2$ the expression can be simplified.

3. Computing the statistics this way is not the same as computing the difference between means (see Table 20-2). While $\bar{D} = \bar{X}_2 - \bar{X}_1$, S_d, the standard deviation of the differences, is not the same as the pooled standard deviations of the populations. This is particularly true when the differences are taken between paired observations. If there are external factors that affect each member of the pair equally, then the standard deviation of the differences will be much smaller than the population standard deviations. In fact, this is the reason for using this form of computation. An example where this would apply would be before/after measurements, such as determining moisture by drying to constant weight.

4. The difference between the two cases is in the computation of s^2. When μ is known, then $s^2 = \sum(X - \mu)^2/n$. When μ is unknown then $s^2 = \sum(X - \bar{X})^2/(n - 1)$.

TABLE 20-2

Hypothetical data showing the relationship between the standard deviation of difference (SDD), the pooled standard deviation of the data and the difference between means, as pointed out in Note 3 of Table 20-1. These synthetic data is in the form of a hypothetical experiment where 10 aliquots have been dried to constant weight.

Sample number	Weight before drying		Weight after drying	Weight difference
1	10.587		9.563	1.024
2	10.974		9.973	1.001
3	11.1		10.089	1.011
4	10.124		9.131	0.993
5	10.254		9.247	1.007
6	10.636		9.652	0.984
7	10.46		9.485	0.975
8	10.342		9.385	0.957
9	10.827		9.842	0.985
10	10.729		9.718	1.011
Mean	10.603		9.608	0.9948 (Mean difference)
Standard deviation	0.314		0.306	0.020 (SDD)
Pooled S.D.		0.31		

The SDD gives a measure of the accuracy of the procedure, while the pooled standard deviation simply represents the variability of the weight of the aliquots being dried. The mean difference is the best estimate of the weight loss for the average size sample.

Exercise for the reader: While computing the SDD is, in general, a good way to determine the accuracy of data from paired observations, in this particular case of moisture determination it is actually a poor way of determining the moisture in the material. Can you tell why? What would be a better way of dealing with these data?

Lessons

There are some important lessons to be gained from the above list and from Table 20-1. Study the entries and note the similarities and differences. The first lesson comes simply from observing the proliferation of cases. The point of this lesson is that seemingly small differences in the formulation of the statistical experiment result in differences in the way the data should be handled. We have pointed this out for particular cases in earlier chapters. Sometimes only one formulation is right, and any other is wrong. However, that is not always the case; sometimes the different statistical hypotheses may extract different information from the data.

We will have more to say about misuses of statistics, but will note here that confusion over the correct hypothesis to formulate in particular situations is probably one of the leading causes of *honest* misuse. After all, with just the few statistics we have discussed so far (means and standard deviations), look at the possibilities, of which all combinations can exist:

(1) population means known or unknown;
(2) population means the same or different;
(3) population standard deviation known or unknown;
(4) population standard deviations the same or different;
(5) same or different numbers of observations;
(6) one-tailed or two-tailed test;
(7) continuous or discrete distributions.

$2^7 = 128$ possible situations!

The importance of hypothesis testing will become more evident when we begin to describe fitting various mathematical models to spectroscopic data. If we have collected data (which, being measurements, are subject to error/variability) and we create a scatter plot of the optical data (absorbance, transmittance, energy, etc.) versus constituent concentration, we can begin to evaluate the fit of the various calibration models to those data using hypothesis testing.

In describing the fit of the calibration models to data using linear through nth order models, we can describe our fitting models as specific examples of the very general case: $y = f(x)$. Having described y as a function of x we can also describe x as a function of y, and we can calculate the expected value for both absorbance (y) and concentration (x). Then, given x we can calculate the expected value for y and vice versa. We use the concept of hypothesis testing when we match these expected (or calculated) values with the observed values.

We begin by determining how the data are distributed and how well the actual (observed) values agree with the expected (calculated/predicted) values. Then we determine whether or not the repeated measurements on the same sample result in data that are Normally distributed.

Having verified that the data are Normally distributed, we can then go on to perform other test, such as testing "goodness of fit". While we caution against overuse of correlation, one goodness-of-fit test is to use the statistic correlation coefficient to see whether the X data are positively or negatively related to the Y data, or whether there is no relationship. We could then determine whether, for example, a linear model or a quadratic model is the proper one to use for the data at hand. The measured correlation coefficient, designated by R, is itself a statistic that estimates the corresponding population parameter, also called correlation coefficient, but designated by ρ. As a statistic, R is also subject to variability, thus percentage points and confidence limits for this statistic are also found in statistical tables.

In calibration work we are also interested in finding "outliers", sometimes called "bad actors" (Tomas Hirschfeld), "mavericks" (Cuthbert Daniel) or "varmints" (Yosemite Sam). These terms are used to describe "bad" data. Bad, unusual, or atypical data are data which do not truly represent the population for which we are trying to develop a mathematical calibration model. If we retain the unusual data during the development of the calibration model, the model will be less accurate than it should, or possibly even invalidated entirely. Unusual data can result from many sources, including hardware failure, operator failure, malfunctions in processing/manufacturing equipment, or an atypical specimen. Detection of outliers is predicated on the assumption that all "good" data are "good" within the confidence limits of the calibration set. The errors of the calibration set are expected to be Normally distributed; then a Z or a t test can be used to determine whether a given datum belongs to the hypothesized population.

As another example, we may use hypothesis testing to determine the detection limit, or the sensitivity of an analytical procedure. In such a case we are actually testing whether a series of measurements for one sample are different (in a statistical sense) from those obtained from a different sample. This is equivalent to a hypothesis test against the null hypothesis that the data support or deny the contention that two populations have the same mean. For a "detection limit" determination, this includes the further assumption that at least one population mean is zero.

These examples illustrate how hypothesis testing has a central place in the evaluation of everyday measurements, as we continue to explore the mathematics behind those measurements.

To work!

Now it is your turn.

Instead of us posing problems, we decided to take the approach of having you find and formulate your own problems. We have some that are ready-made for you. In earlier chapters we have, from time to time, presented questions that we claimed the science of statistics could answer. Now is the time to go back and look them up. For example, in Chapter 1 we presented a spectrum with a marginally noticeable absorbance band. How would you decide whether the band is real? Since we have not provided data, you will not actually be able to answer this question (or any of the others) directly. However, you *can* do the following:

(1) Put the question into statistical formalism, i.e., write it down in the nine steps that go into the formalism of hypothesis testing: What is the experimental question? What is the statistical question? What is the proper statistic (key question)? What value to use for alpha? One-sided or two-sided? And so on. Only the last few of the nine points will be unanswerable due to lack of data.

(2) Supply your own data for some of these questions.

(3) Create your own questions (you knew that was coming).

(4) Are any (or all) of these questions unanswerable using the statistics we have already discussed (probably so, there are many statistics we have not yet covered)?
(5) What problems in your laboratory can you now attack that you did not know how to before?
(6) What problems in your laboratory can you now attack that you did not think of before (perhaps problems that you only attacked qualitatively but felt you could not deal with quantitatively because "there is too much variability")?

There is still a good deal of material to cover. Some of this will help you answer the above questions, some will open up whole new areas of questions that statistics can help answer. We begin this expansion in Chapter 21, where we learn how to count statistically.

References

20-1. Dixon, W.J., and Massey, F.J., *Introduction to Statistical Analysis*, 4th ed., McGraw-Hill, New York, 1983.

21

How to Count

In Chapter 19 one of the items discussed (not at too great length) was the question of doing hypothesis testing on proportions. A proportion looks like other data, but in fact is different. The reason proportions look the same is that they are expressed as decimal fractions (between 0 and 1). The difference is that, unlike measured quantities that (in principle at least) can have an infinite number of possible values in any given range, proportions fall into an entirely different class of measurement. This new measurement category is the one that deals with discrete objects that are counted in order to produce data. Note that since we specifically rule out taking measurements on the objects, the only characteristic we can note is whether or not a given object has a certain property. It is also possible to extend this by noting several characteristics, and divide the objects into more than two categories. If the object has the property, we assign it to one class, if not we assign it to another class. Then we count the objects in each class. When objects are counted, the important factors are the total number of objects involved, and the fraction, or *proportion* of objects that have the given property.

Depending upon the conditions, data that result from counting experiments can follow one of several distributions.

The first case is the one that we have already run into: in Chapter 5 we listed all the combinations of the integers 1–10 taken three at a time (actually we listed the means, but let us not quibble). This list included the number of times each mean occurred during an exhaustive listing of the combinations, and subsequently we showed that the "bell-shaped curve" we obtained followed (or at least approximated) the Normal distribution. This, we eventually showed, was due to the combinatorial properties of numbers, and ultimately demonstrated that the central limit theorem requires that *any* distribution approximates the Normal distribution under some very general conditions. Thus, even though the list was in reality obtained by counting the objects involved (i.e., the various numbers used) the distribution tended to be Normal. We also showed that the approximation to the Normal distribution got better and better as the number of numbers (hate that, but cannot think of anything better) increased.

Well, in those discussions, we showed what the distribution was going *to*, but not what it was coming *from*. Since *any* distribution approximates Normality when certain conditions hold, when those conditions *do not* hold the distribution reverts to the original, underlying one. The point at which the distribution changes is not a sharp one. Rather, the choice of which distribution to use (the "true" one or the Normal approximation) depends upon how good the Normal approximation is in a given case, and also on how critical a factor the difference makes. This last criterion is a non-statistical question, so formal statistics cannot help.

Remembering what we said just above, when counting, the important factors are the total number of objects involved, and the fraction, or *proportion* of objects that have the given property. The property that allows the sampling distributions of counted objects to approximate Normality is that there are a lot of numbers involved, and the proportions are high. If either of these conditions is not met, then the distribution will not be Normal. The distribution that will actually be obtained under those circumstances will depend upon which condition was not met.

The actual distribution that gave rise to the tables in Chapter 5, as well as the discussions in Chapters 3, 4, and 11, is the multinomial distribution. More familiar, perhaps, is the binomial distribution, which is a special case of the multinomial distribution. In the multinomial case, we computed means from numbers that could range from 1 to 10 (decanomial?) with equal probability of being selected from the population. If, on the other hand, we do the same calculation for numbers that can only have values zero and unity (or one and two, or any other pair), each of which can be selected with equal probability, the distribution will be binomial.

We modified the program we used and presented previously to limit the range of possible values to 5 and 6. This was done simply by replacing the population 1 – 10 with a new population: 5,6,5,6,5,6,5,6,5,6. These elements were selected so that the population mean, 5.5, is the same as the population mean that we had before. We also modified the program to write the results directly out to a disk file, for the sake of convenience. The modified program is shown in Table 21-1. Do not forget to change statement 35 as well as statement 10 when trying different values of n.

TABLE 21-1
BASIC program to compute random members from the binomial distribution.

```
10 N = 1
15 DIM POP(10)
20 REM randomize timer :REM remove first remark to change numbers
30 FOR I = 1 TO 10: READ POP(I): NEXT I: REM get data (from line 450)
35 OPEN "O", #1, "N1"
40 FOR I = 1 TO 20 : REM generate 20 averages
50 GOSUB 9000 :REM get an average of n numbers
300 PRINT#1, XBAR :REM print it
350 NEXT I : REM end of loop
360 CLOSE #1
400 END : REM end of program
450 DATA 5,6,5,6,5,6,5,6,5,6
9000 REM
9001 REM subroutine to average n numbers from POP
9002 REM
9050 SUMX = 0 : REM initialize
9090 FOR J = 1 TO N :REM we're going to average n numbers
9100 GOSUB 10000 : REM get a random number
9120 SUMX = SUMX + A : REM generate the sum
9130 NEXT J :REM end of summation
9140 XBAR = SUMX/N : REM calculate mean
9150 RETURN
10000 REM
10001 REM subroutine to get a random population member
10002 REM
10010 B = RND(B) :REM generate random number
10020 A = INT(10*B) + 1: REM convert to an index
10030 A = POP(A) :REM get a member of the population
10040 RETURN
```

Table 21-2 shows the results of running this revised program, for several values of n, the number of numbers that contribute to the mean. We arranged it so as to cover the same cases as we did for the population of integers 1 – 10, so that the tables can be compared on a one-for-one basis. Compare this table with the previous one. What are the differences? The similarities? What are the standard deviations of each set of 20 samples from this population? (*This* population? Which population is being referred to?) Do they also decrease inversely with $n^{1/2}$?

As an exercise, reread all the pertinent chapters (3, 4, 5, and 11) and repeat all the computations done there, particularly the exhaustive listing of $\bar{X}$, S, and t, using the binomial parent distribution. (This should be easy if you did them the first time, as we recommended: the required changes to the program are trivial—mainly a matter of changing the population, the same way we did here). Inspect

TABLE 21-2

Each column contains 20 means of random samples from a population of integers (5 and 6), where n samples were used to calculate each mean. Note that the population mean is 5.5.

$n = 1$	$n = 2$	$n = 5$	$n = 10$	$n = 20$	$n = 50$	$n = 100$
6	6	5.6	5.5	5.7	5.64	5.58
6	5	5.4	5.9	5.55	5.52	5.45
5	6	6	5.6	5.6	5.42	5.47
5	5	5.8	5.5	5.55	5.48	5.62
6	5.5	5.6	5.7	5.5	5.4	5.55
6	6	5.6	5.5	5.55	5.54	5.55
5	6	5.8	5.5	5.35	5.62	5.52
5	6	5.2	5.6	5.45	5.62	5.54
5	6	5.6	5.6	5.4	5.62	5.47
6	5.5	5.8	5.4	5.5	5.48	5.51
6	6	5.4	5.6	5.3	5.64	5.45
6	5	5.6	5.5	5.45	5.46	5.52
6	6	5.4	5.5	5.55	5.48	5.51
6	6	5.6	5.2	5.55	5.56	5.45
6	5	5.4	5.3	5.5	5.56	5.54
6	6	5.8	5.6	5.65	5.52	5.5
6	6	5.6	5.4	5.55	5.44	5.48
6	5	5.6	5.4	5.7	5.5	5.61
6	5.5	5.6	5.5	5.6	5.48	5.4
5	5	5.2	5.5	5.6	5.54	5.55

the results in the light of what we have been talking about in this book. Again, where do the results differ between the two cases and where are they the same? The answers to these questions are not only of theoretical interest. It may become important to decide whether a given situation must be treated exactly, using the correct distribution (assuming it can be determined) or whether using the Normal distribution will be "good enough". In such a situation, to make the decision will require knowing not only how critical the difference will be, but also knowing how good the Normal approximation is under different conditions.

To start you off, we will discuss the first couple of questions above. First of all, note that for the case $n = 1$, the elements are simply random selections from the initial population. Therefore, they can only range from 5 to 6 in the binomial case, and from 1 to 10 in the multinomial case. In both cases, however, the samples contain a uniform distribution of values, because the population contains a uniform distribution of values. The range of values in the binomial case remains smaller than for the multinomial case for the various values of n used. The number of discrete values is also smaller in the binomial case.

Twenty samples is far too small a number to determine the distribution of samples very well, but if you have done the exhaustive computations we

TABLE 21-3
The beginning of Pascal's triangle. The numbers give the relative frequency of occurrence of the various combinations of elements from the parent population.

$n = 0$						1					
$n = 1$					1		1				
$n = 2$				1		2		1			
$n = 3$			1		3		3		1		
$n = 4$		1		4		6		4		1	
$n = 5$	1		5		10		10		5		1

recommended above you should be able to answer the following questions, which depend upon the distribution of the samples: does the range remain smaller? We noted previously that the distribution of samples in the multinomial case forms a peak, which, by virtue of the central limit theorem becomes Normal. The binomial distribution also becomes Normal when n is large enough. Which one approximates Normality faster? Which one approximates the t distribution faster when dividing $\bar{X}$ by S?

The theoretical binomial distribution follows Pascal's triangle [see Table 21-3—and also the interesting and excellent discussion of Pascal's triangle in the classic work by Knuth (see Volume 1: "Fundamental Algorithms", pp. 51–68 in Reference (21-1)), and the original treatise by Pascal (21-2)]. The probability of finding any given value, when computing the mean of a sample, is given by the corresponding entry in the triangle, divided by the sum of the entries in the corresponding row. The entries in Table 21-3 show that the binomial distribution also has a maximum at the center of the distribution, thus this distribution is the third one we have come across that has a "bell-shaped curve". (Exercise for the reader: what are the other two?)

Another characteristic of Pascal's triangle to note is that the sum of each row is equal to 2^n. Thus, for example, for $n = 3$ the probability of obtaining 5.3333... by random selection is 3/8.

By this reasoning, the column of Table 21-2 corresponding to $n = 2$ should have 10 entries equal to 5.5, yet there are only three. Can you tell why? Does this indicate that the binomial distribution does not hold as we have been saying? Will Alice marry John?

Do your homework. We will continue in the next chapter with some examples relating all this to chemical measurements.

References

21-1. Knuth, D.E., *The Art of Computer Programming*, Addison-Wesley, Menlo Park, CA, 1981.

21-2. Pascal, B., "Treatise on the Arithmetical Triangle", Hutchins, R.M. ed., *Great Books of the Western World*, Vol. 33. University of Chicago Press, Chicago, IL, 1982, pp. 447–473.

22

AND STILL COUNTING

The binomial distribution arises when the data consist of objects that are counted, and the value reported is the fraction, or proportion, meeting some qualitative criterion. In this case, when there are many objects, and a fairly large proportion of them meets the necessary criterion, the distribution will be approximately Normal and the following expression can be used to perform hypothesis testing:

$$Z = \frac{\bar{X} - \mu_0 \pm (1/2N)}{(P_0(1 - P_0)/N)^{1/2}} \tag{22-1}$$

where $\bar{X}$ is the fraction of the measured data having the property in question and P_0, the population proportion, replaces μ as hypothesized population mean. Contrary to the usual rule about nomenclature, P is used to indicate the population proportion even though it is not a Greek letter.

How many are "many"? In critical cases more exact formulas are available, but the rule of thumb is that the Normal approximation will suffice if $NP > 5$ (note that NP, which we shall see often, is the count of the items meeting the given criterion). This makes some intuitive sense. The Normal distribution is continuous, whereas the binomial distribution, being based on counting, is discrete. In order for the Normal approximation to hold, there must be enough cases involved (i.e., NP should be large enough) that the distance from one discrete value to the next should be unnoticeable, and therefore act as though it were, in fact, at least quasi-continuous.

When $NP < 5$ then more accurate (and complicated) expressions, based on the binomial distribution should be used. This sampling distribution is given by the expression:

$$\frac{X}{N} = \frac{N!}{X!(N-X)!} P^X (1-P)^{N-X} \tag{22-2}$$

where P represents the actual fraction (proportion, or probability of finding) of items in the population, and X items will be found in a sample of size N.

Since X and N are discrete, it is actually possible to enumerate the values of X/N corresponding to the various values of X, something we could not do for continuous distributions, such as the Normal and t distributions. To determine the probability points, however, it is still necessary to add up the values for all the different cases, in order to ascertain the range of values for X/N that will be found 95% (or 99%, or 97.5%, or whichever confidence level is needed) of the time when the true proportion of the population is P. Then hypothesis tests can be conducted in the same manner as when the continuous distributions are used.

The book by Dixon and Massey (22-1), which we have recommended before, has a fairly extensive discussion about when to use which distribution, and the necessary formulas.

All the properties of the binomial distribution for a given situation are determined by the values of N and P for that situation. In particular, the values for the mean and variance of the sampling distribution for samples of size N from a population containing a proportion P of items are determined by these parameters. As it happens, the formulas for the expected values of these statistics are particularly simple when expressed this way. Thus:

The mean: $\mathscr{M} = \mathrm{Np}$

The variance: $\mathscr{V} = \mathrm{Np}(1-p)$

This also makes intuitive sense. Since P represents the fraction of the time that the "event" occurs, the larger the sample we take the larger the number of "hits" we should get.

In the foregoing discussions, we have been assuming that P represents the proportion of data exhibiting a characteristic. It is clear from probability theory, however, that if this is the case then the proportion of cases *not* having that characteristic is $(1-P)$. It is a short step from this to the realization that it is only a matter of definition as to which situation represents the "event" being counted. There is a symmetry here. This symmetry even shows up in the formulas that result when the binomial distribution is used: there is invariably a term of the form $P(1-P)$, as there is in Equation 22-1. More importantly, this must be kept in mind when deciding which distribution to apply in any given case. For example: since, as we have discussed earlier, in order to apply the Normal distribution approximation the product NP must be greater than 5, by this symmetry the product $N(1 - P)$ must also be greater than 5, otherwise the approximation will not be valid.

The binomial distribution must be used if N is small regardless of the value of P. Usually, N is not considered by itself, the product NP is used as the factor in making these decisions. However, when N (or NP) is large but P is small, then the binomial distribution grades into a still different distribution, the Poisson distribution. The Poisson distribution generally applies when very rare events are being counted, so that only a few cases are available. Radioactive decays, for example, follow this distribution, as do photon-counting statistics in visible/UV spectroscopy (Table 22-1).

The Poisson distribution is given by the formula:

$$\frac{X}{N} = \frac{e^{-NP}(NP)^X}{X!} \tag{22-3}$$

which, unlike most statistics, is relatively straightforward to evaluate, and thereby generate the probability tables. This distribution also has some peculiar characteristics, the most interesting of which is perhaps the fact that not only is the expected value of the mean equal to NP (like the binomial distribution), but so is the expected value of the variance. Thus, for any quantity distributed according to the Poisson distribution, the standard deviation is equal to the square root of the value.

Indeed, the quantity NP is so pervasive in discussions of the Poisson distribution that NP is itself often considered the parameter of the distribution (usually λ or ξ is used for this parameter—the convention of using Greek letters holds for the Poisson distribution, even though not for the binomial), and all quantities referred to it, rather than to P alone. This point is not trivial. Interpretation of tables of percentage points for the Poisson distribution may be confusing or impossible if the point is not kept in mind.

For example, when the tabled value of $\xi = 0.1$, this may mean that $P = 0.01$ while $N = 10$, or it may mean that $p = 0.001$ and $N = 100$.

There are some strange effects that occur, which we can use to illustrate the application of this concept. For example, suppose a (random) sample is taken

TABLE 22-1

Distributions for various combinations of P and NP.

	NP	
	Large	Small
Large *P*	Normal	Binomial
Small *P*	Poisson	Binomial

from a population in which it is suspected that there may be some quantity of interest. This might be a jar of supposedly pure optically active crystals which is suspected of being contaminated with crystals of the opposite rotation (presumably well mixed in). A sample of size N is taken and each crystal analyzed for the direction of its optical rotation. No defective crystals are found. Does this prove the jar is pure?

We will forgo the formalism of the hypothesis test, and simply note here that the 95% confidence limits for $\bar{X} = 0$ are (from tables): LCL: $\xi = 0$ and UCL: $\xi = 3.68$. Now, if ξ truly equals zero then the jar is pure (mathematically, $\xi = NP = 0$, therefore $p = 0/N$, and since N is non-zero, $p = 0$), but we do not know this, we only know $\bar{X} = 0$ (there might be some impurity crystals present, that we missed). The constant of the distribution, which is what is tabled, still does not tell us whether there is contamination or not, this value must be used in conjunction with the number of crystals analyzed. Since the UCL is 3.68, we must solve for P for various values of N, the number of crystals measured. Thus, since $P = \xi/N$, the more crystals we measure, the more confidence we have that the jar is pure, but we can never be sure. If we measure 10 crystals (and presumably do not find any defects: as soon as we find even one crystal with the wrong rotation we have proven that the jar is contaminated), we only know that the proportion of defective crystals is (with 95% probability) less than $\xi/N = 3.6/10 = 0.36$. Thus, if only 10 crystals are measured, then the possibility exists (with a 5% probability) that over 1/3 of the crystals could be bad and we would not have detected the fact!

We can trade off assurance of purity against the amount of work needed. If we measure more crystals, we guard against ever-decreasing amounts of potential contamination. Since the product NP is constant, by increasing N we reduce the value of P corresponding to the UCL of ξ. Clearly, if we take 50 crystals, for example, then we do not find any defects we are confident that the proportion of impurity is less than 3.6/50; or there must be less than 7% bad.

This is another example of what we have pointed out before: statistics can disprove, but never prove. A single bad crystal disproves the null hypothesis $\xi = 0$, but we can never prove it, we can only push the possible levels of contamination down to lower and lower limits, by taking larger and larger samples.

Another interesting, albeit more prosaic application comes from our own experience. While attempting to calibrate an energy-dispersive X-ray fluorescence spectrometer to measure the salt (NaCl) content of a group of specimens containing approximately 1% salt, we found it impossible to obtain a precision of better than 0.1%, or 10% relative. After tearing our hair for a suitable length of time because we could not find anything wrong with the equipment, we thought to check the sampling distribution of the salt in the samples.

A 1–2 g sample (our best estimate of the amount of sample actively fluorescing) would contain 10–20 mg of salt. Checking the samples with a microscope revealed that the size of the salt particles was approximately 0.5 mm, giving a volume of 0.12×10^{-3} cm^3. The density of salt is just about 2 g/cm^3, thus each crystal contains about 0.24 mg of salt. Comparing this value with the 10–20 mg of salt in each sample, we find that there are only 40–80 salt particles in each sample. As we have pointed out previously, the standard deviation of a Poisson-distributed quantity is the square root of the counts; the square root of 40 is 6.3, and of 80 is 9, therefore the relative variation is between 6/40 and 9/80 (= 0.15 and 0.11). This is close enough to the 10% relative variation we found in our precision data to lead us to the conclusion that the sampling variation of the salt particles in our specimens was the cause of our difficulties.

References

22-1. Dixon, W.J., and Massey, F.J., *Introduction to Statistical Analysis*, 4th ed., McGraw-Hill, New York, 1983.

23

CONTINGENCY TABLES

In Chapters 21 and 22 we discussed several of the distributions that arise when the quantity being used for computation is the proportion, or relative frequency, of items meeting (or failing to meet) a given criterion rather than being the result of a measurement of a continuous physical property. The distributions encountered so far were the Normal, binomial, and Poisson distributions.

There is one more distribution that is of importance when dealing with proportions; indeed, this distribution is perhaps of more importance than all the others together due to its wide range of applicability, and no discussion dealing with the statistics of proportions would be complete (some would say not even viable) without it. The distribution is one that we have met before: the χ^2 distribution. The χ^2 distribution is of supreme importance when objects (or observations) are categorized based on the proportion meeting one of several criteria.

We will discuss some of the details below, but first a diatribe on why this is so important. We will repeat, *ad nauseum* if necessary, what statistics is all about. Statistics is the science that allows us to determine whether an effect exists when the data that purport to measure the effect are contaminated with randomness. The label we apply to the random part of the data, whether we call it "noise", "error", or something else is of no consequence. The point is that in order to verify that an effect exists, it is necessary to determine that the data could not have arisen solely by the action of the random factor alone, and we have spent considerable time and effort in presenting some of the means by which the science of statistics allows this determination to be made in an objective manner (hypothesis testing, and all that). All hypothesis tests, however, are based on certain assumptions. For the cases we have studied this is, first, the assumption that the data observed were obtained from a single random sample of some population. The population characteristics may or may not be specified, but what is specified, in all cases, is the nature of the random factor with which the data are contaminated. The random factor is also assumed to come from some population, but for this population certain characteristics are always specified.

Some assumptions about the nature of the random factor apply only to certain hypothesis tests and not to others, for example, whether the random factor (noise) is Normally distributed.

There are two characteristics, however, that are always assumed. These characteristics are first, that the noise is, in fact, random (in the sense we have previously described in Chapter 2), and second, that the contribution of the noise to any given measurement is independent of the noise contribution to any other measurement.

We do not wish to dwell on the question of defining these two key characteristics here, although we will return to that topic. What is critical at this point, however, is to appreciate the fact that all the hypothesis testing, and all the confidence interval and confidence level discussions we have presented until now are based on the assumption that these characteristics (that the errors are random and independent) are always found in the data.

It is all very well to go ahead and do a hypothesis test, find that the calculated statistic is beyond the critical value and conclude that the null hypothesis should be rejected and that therefore a real physical effect is operative.

The careful statistician (and certainly the *experienced* statistician, i.e., any statistician who has been bitten by this mistake) knows better. What he knows better is this: that if the null hypothesis has been rejected, then he can conclude that a real physical effect exists *if and only if the assumptions upon which the hypothesis tests are based, and for which the tables of critical values of the various statistics are calculated, hold for the data he is doing his calculations for.* He knows that otherwise, the computation of a value for a statistic outside the specified confidence interval could be due equally well to the failure of the assumptions to hold as to the presence of the physical effect sought. He knows that failure to verify that these fundamental assumptions hold for the data at hand constitute the second most common source of honest misuse of statistics. (Exercise for the reader: what did we claim is the most common?)

Thus it is, that before jumping to conclusions, any statistician worth his salt will test these assumptions. Indeed, a good statistician will test the assumptions *first*, and if the data prove defective because they fail *these* tests, would not dream of testing the main hypotheses until the defects have been corrected. Sometimes the defects can be removed by manipulating the data itself; e.g., outliers can be removed from the dataset before analysis. Sometimes, however, new data must be collected, either to supplement the data already available, or using a modified experimental design, or taking into account phenomena previously ignored (perhaps serendipity has occurred!), or in extreme cases, the original experiment was so badly botched that an entirely new experiment must be devised and performed.

The point, however, is that it does not suffice to carelessly collect data, perform a hypothesis test and then claim that the results must be right because "it was checked statistically". A careful scientist will cheerfully (well, maybe not *so* cheerfully, but he will do it) redo an experiment when he realizes that there is a defect in the physical setup of the experiment that might invalidate the data. He must be equally critical of experiments where the data are defective, even if the physical cause is not so obvious.

End of diatribe.

So what has our diatribe to do with our earlier claim that the χ^2 statistic is one of supreme importance? Putting two and two together should make the answer obvious by now: it will help us determine whether or not a given set of data does, in fact, meet the necessary criteria.

It turns out that the variances of random samples from Normal distributions constitute only one of the cases that follow the χ^2 distribution. There are actually several other situations where data follow this distribution, and the one of interest to us here and now is that the variability of proportions is one of these "other situations". As we have seen before, when objects are classified into two categories, then the sampling distribution can be Normal, binomial, or Poisson. However, if the data can be classified into two or more categories, then the following relation holds:

$$\chi^2 = \sum_{i=1}^{k} \frac{(o_i - e_i)^2}{e_i} \tag{23-1}$$

In Equation 23-1, o_i represents the observed fraction of the data in the ith category, while e_i represents the fraction of the data *expected* to fall into the ith category. This computed result is χ^2 distributed with k degrees of freedom. Two very good discussions of this aspect of the χ^2 distribution can be found in References (23-1,23-2).

The variability of the data around the expected value is probably the single most powerful tool available for performing these tests. There are other tools that are better suited to particular cases, but this use of χ^2 is the one that is simultaneously of most power and generality, because it can be used to test not only randomness and independence, but also with proper organization it can be used to test the distribution of the data (or of the errors).

To see how these tests work, let us take a very simplistic situation. There is a table of 100 numbers (see Table 23-1A), which are purported to be random samples from a uniform distribution of numbers from 1 to 100. We wish to test the claim that these numbers were actually selected at random.

To perform the test, we note that if the population is uniformly distributed, then there should be approximately an equal number of them in any set of uniform intervals. It is convenient to take 10 intervals. This number of intervals provides a test with a reasonable number of degrees of freedom, while at the same time giving a reasonable amount of numbers in each range. If the numbers are indeed random, then their variability should follow the χ^2 distribution according to Equation 23-1. A requirement of performing χ^2 tests on proportion data is that each group into which the data is to be classified must have a minimum expected value of five members. Clearly, if 100 items are to be distributed uniformly among 10 categories, then the expected number in each category is 10, so this requirement is satisfied.

Table 23-1B shows the computations. We can now perform a hypothesis test, which we do according to the proper formalism:

(1) We state the experimental question: Are the data of Table 23-1A random samples from a uniform distribution?
(2) The null hypothesis is Ho: $\chi^2 = 10$. The alternative hypothesis is Ha: $\chi^2 \neq 10$.
(3) We will use a level of significance $\alpha = 0.05$.
(4) The test statistic is:

$$\chi^2 = \sum_{i=1}^{k} \frac{(o_i - e_i)^2}{e_i}.$$

(5) The critical values are: $2.70 < \chi^2 < 19.02$.

TABLE 23-1A
Data to be tested for randomness.

69	90	88	12	83	78	76	80	27	100
41	84	52	71	85	49	3	43	8	44
96	74	14	82	89	32	47	94	65	20
95	15	70	92	86	57	31	17	93	72
68	36	16	56	25	81	54	30	18	10
50	7	60	79	97	42	73	29	23	2
53	87	91	64	77	98	46	40	48	59
21	11	33	99	35	37	4	24	58	13
6	5	19	66	45	9	1	28	62	34
75	26	51	67	63	61	38	22	39	55

TABLE 23-1B
Computations of χ^2 for the data given in Table 23-1A.

Range	Observed	Expected	$(o - e)$	$(o - e)^2$
1–10	10	10	0	0
11–20	10	10	0	0
21–30	10	10	0	0
31–40	10	10	0	0
41–50	10	10	0	0
51–60	10	10	0	0
61–70	10	10	0	0
71–80	10	10	0	0
81–90	10	10	0	0
91–100	10	10	0	0
				$\sum\frac{(o - e)^2}{e} = 0$

(6) Collect the data (see Table 23-1).
(7) Calculate the test statistic: $\chi^2 = 0$.
(8) State the statistical conclusion: The calculated value is outside the confidence interval for the statistic, and therefore the measured value of χ^2 is statistically significant.
(9) State the experimental conclusion: the data in Table 23-1 are not a random sample from a uniform distribution of numbers.

Before we continue our examination of this hypothesis test, we pause for a note (C-sharp): if you followed along as we did the hypothesis test, you would have seen that we used the confidence limits for the χ^2 statistic for nine degrees of freedom. This is because, when performing hypothesis tests on goodness-of-fit data (as this test is), the degrees of freedom are one less than the number of categories into which the data are grouped. Here we will only note that this is the way it is done; see Reference (23-1) for more details about this.

Because of the limitations of time and space, we often try to pack more than one lesson into some of our examples, and this is one such case. While we have illustrated the use of the computation of χ^2, as we planned to do, this was only the surface lesson of the exercise. The actual result of the hypothesis test contains at least two more lessons. First, note the statement of the conclusion: Table 23-1 is not a random sample from a uniform distribution. Note very clearly that this only tells us what Table 23-1A is *not*, there is no information as to what Table 23-1 *is*.

However, for the second "extra" lesson of the exercise we can glean some information from the way in which the data were rejected. Note that the null hypothesis was rejected because the computed value of χ^2 is too small for the null hypothesis to be accepted. This means that the observed values of the frequencies within each range (category) were too close to the expected values. In other words, while there seemed to be the right amount of numbers in each range, the amount of numbers in each range was *too* right, the fit of the numbers was too perfect, for it to have been random.

We will have more to say about randomness in later chapters too, but if you have learned from this example that the *appearance* of randomness can be deceiving, you have earned an A + for this lesson.

The heading of this chapter is "Contingency Tables"; where does that fit in? In the fonts of statistical wisdom that we have recommended (23-1,23-2) are extensive discussions of contingency tables. Basically, a contingency table is simply a two-way layout of the number of items in the sample set that fall into different categories. The difference between this layout and our previous list of categories is that, in the contingency table, the horizontal and vertical axes represent different types of categorization. Actually, these can represent *any* type of category that might be important in a given experiment. The importance of this layout to our current discussion is that it is a way of testing *independence* of data. This is due to the effect of the layout, not on the actual entry in each position, but on the expected value for each entry. In a table with two columns, for example, if one column represents a case that should happen 4/5 of the time and the other column represents a case that should happen 1/5 of the time, then in every row, the first column should contain 4/5 of the total entries for that row and the other column 1/5 of the total entries. If the row and columns represent factors that are indeed actually independent, then these relationships will hold regardless of whatever else is going on in the data.

So if, in fact, the two axes represent independent ways of categorizing the data, then the observed values will follow that scheme, and can then be tested with the χ^2 statistic. This, then, is a way of testing the *independence* of the variations of a set of data, separate and distinct from the test of randomness. For example, half the data in Table 23-1A represents values in the range 1–50, and the other half values in the range 51–100. If the values in the table are independent, then there should be no relationship between any number in the table and the one following it. In other words, if the nth number in the table is between 1 and 50, then the $n+1$th entry should have a 50% chance of also being in that range, and 50% chance of being in the range 51–100. Since half of the numbers will be in the range 1–50, then the expected value for the $n+1$th being in either of the subranges will be 25; similarly, there should be 25 numbers in each range following a number in the range 51–100. Table 23-2A shows the contingency table for this situation, and Table 23-2B presents the calculation of χ^2 for this contingency table. Forgoing the formalism (exercise for the reader: redo this hypothesis test using the formalism), we note that the 95% confidence interval for

TABLE 23-2A

Contingency table for data of Table 23-1A. The entries in this table have been labeled with Roman numerals for convenience in referring to them.

	$N + 1$th entry < 50.5	$N + 1$th entry > 50.5
Nth entry < 50.5	I = 27	II = 22
Nth entry > 50.5	III = 22	IV = 28

χ^2 for one degree of freedom is $0.00098 < \chi^2 < 9.35$. The value calculated from the data, 1.22, is comfortably within that range; therefore, we conclude that the data show no evidence for any relationship between successive numbers: each entry is independent of the preceding one.

Another note here regarding the hypothesis test on contingency tables: why did we do the test using only one degree of freedom to determine the critical values? Well, you sort of have to take our word for it (or Dixon and Massey's (23-1)—that is a very reliable word—or some other statistician's word) but the number of degrees of freedom in a contingency table is computed as $(r - 1)(c - 1)$ where r and c are the number of rows and columns, respectively, in the contingency table. Since this particular table has two rows and two columns, the number of degrees of freedom $= (2 - 1)(2 - 1) = 1$.

There are three more important lessons from this contingency table. First, we have tested for only one of many possible relationships that might exist among the numbers. For practice in setting up contingency tables and calculating χ^2 for them, test for the following possible relationships:

(1) Is there a relationship between the nth number being in one of the three possible uniformly spaced intervals (i.e., 1–33, 34–66, 67–100) and the $n + 1$th being above or below 50.5?

TABLE 23-2B

Computations of χ^2 for the data given in Table 23-2A.

Entry	Observed	Expected	$(o - e)$	$(o - e)^2$	$(o - e)^2/e$
I	27	24.5	2.5	6.25	0.25
II	22	24.5	−2.5	6.25	0.25
III	22	25	−3	9	0.36
IV	28	25	3	9	0.36
					$\sum \frac{(o - e)^2}{e} = 1.22$

(2) Is there a relationship between the nth number being above or below 50.5 and the $n + 1$th being in the lowest quartile?

Note that there are innumerable possible combinations of relationships that can be tested for, since not only can the range be split into a number of intervals, but also there is no requirement that the intervals be the same. The requirement is that, if they are not the same, the proper expected value be used for each entry in the contingency table. For example, in exercise #2 above, there will be two expected values of 12.5, and two of 37.5. The only rule here is the same as before: The table must be set up so that the minimum expected value for each entry is at least five. Thus, for the data of Table 23-1A that we are using, the largest set of possible relationships that we can test is 4×5 intervals. Other potential relationships exist between each number and the second one after it, the third, etc.

The second important lesson is that we have demonstrated that, even though the data were found not to be randomly selected, the entries are independent; it is important not to confuse the two concepts. In fact, since we generated the data, we know exactly how they came about: the full list of numbers from 1–100 was selected; this selection was indeed not random. The numbers were then scrambled, using a random number table. Thus, the entry in any position was indeed selected independently of the number in any other position.

The third important lesson is that having tested for (and not found) one relationship, we have not ruled out other possible relationships that could exist in the data. This has important consequences when we pursue the subject of testing for randomness and independence, which we will do in the next chapter.

References

23-1. Dixon, W.J., and Massey, F.J., *Introduction to Statistical Analysis*, 4th ed., McGraw-Hill, New York, 1983.

23-2. Alder, H.L., and Roessler, E.B., *Introduction to Probability and Statistics*, 5th ed., W.H. Freeman and Co., San Francisco, 1972.

24

WHAT DO YOU MEAN: RANDOM?

As we pointed out in Chapter 23, one of the most important factors to verify about data that are to be subjected to statistical testing is that, in fact, the errors that are supposed to be random actually are. We also pointed out that it is not sufficient to assume that the error structure of a given set of data is random; this assumption must be tested.

This gives rise to two problems: first, how to adequately test whether the errors of any given set of data are actually random, and second, what does it mean to say that a given set of values is random, anyway? Indeed, the second problem is an outgrowth of the first, because it is difficult to test for something that is not even defined.

Now, the best discussion of randomness that we know of is the one by Knuth (24-1)—but then, Knuth's discussion of *anything* related to computers is probably one of the best available. Knuth couches his discussion in terms of generating random numbers using a computer, but it is applicable to any numbers that are purported to be random.

Knuth spends fully 35 pages of his classic work just on adequately defining random numbers; it is a particularly slippery concept. We could not do justice here to either the concept or to Knuth's treatment if we were to try to, so all we can hope to do is to nibble around the edges and try to give an idea of what is involved.

Knuth actually presents (or at least proposes) a number of definitions, and shows that most of them are either too strong, in that no possible set of numbers could fulfill the definition, or too weak, because it is possible to construct sets of numbers that meet the definition yet are obviously and demonstrably non-random. Unfortunately, most of Knuth's definitions are incomprehensible without having read the entire discussion so as to have the background of the terminology he uses, but perhaps we can give the flavor of the nature of these definitions by discussing one of the more comprehensible ones.

The discussion begins by noting that it is not possible to determine if a single, isolated number—say, 0.179632—is random or not: it depends upon how the number was arrived at, and how it stands in relation to other numbers. Numbers can be said to be random or non-random only when sequences are under consideration.

A sequence of numbers can be made into a single number. One simple way to do this is to multiply each number in the sequence by a suitable constant, then add the results. For example, the numbers in the sequence 3,5,1,8,2 can be made into the single number 0.35182 by multiplying each member of the sequence by 10^{-k} (where k is the position of the number in the sequence) and then adding these results.

Having created a single number, Knuth then notes that the value of the number can be expressed in any number base system whatsoever. This step is necessary to insure that the results arrived at are completely general and do not depend upon the number system in use. Therefore, he assumes that the number generated is expressed in the number base b (where b can have any value: in a different chapter he discusses number systems using bases that are negative, fractional, irrational, and other strange sorts of number bases). Furthermore, the number is scaled (if necessary) so that it represents a fraction between 0 and 1. The number, which is now $x_1 b^{-1} + x_2 b^{-2} + \cdots$ is thus represented as x_1, x_2,... The randomness of the original sequence has now been transformed into the randomness of the sequence of digits of the number generated from the original sequence. Then the definition of "random" for this case is that a sequence of numbers is random if:

$$\lim_{k \to \infty} \Pr(X_n \ldots X_{n+k-1} = X_1 \ldots X_k) = 1/b^k$$

In words, this says that for all possible subsequences of length k of the original sequence (where k takes all values between 1 and infinity), the probability of obtaining that particular subsequence of (base-b) digits is inversely proportional to b^k, where k is the length of the subsequence. Since there are b^k possible subsequences of length k, the theoretical probability of any one of them occurring is $1/b^k$, and if all subsequences do in fact occur with appropriate probability, then the sequence must be random.

Knuth shows that for rigorous mathematical applications this particular definition is not completely satisfactory. Nevertheless, it is actually a fairly good definition for more casual uses, and it is particularly good as an instructive definition for us to examine here.

There are several key points for us to note. The first is that the definition is couched in terms of probability. No single number can be said to be random (as we noted before). Conversely, any truly random sequence is certain to contain within it any particular subsequence *somewhere*, if the sequence is long enough. Only a probabilistic definition can cope with these conditions.

The second key point is the fact that the definition specifies the probability for *all possible* subsequences. This means not only the ones that originally went into the creation of the full sequence, but all combinations of those, and also all combinations of segments of the generating numbers such as can arise if the original numbers contained more than one digit. For example, if the numbers 27, 35, 94, 77 ... went into the creation of the sequence 0.27359477..., then among the subsequences to be considered are those that are not simply combinations of the original numbers, such as 7359, 477, etc. as well as those that are, such as 2735, 359477, ...

Another key point is that the definition specifies an infinite sequence. While there are some proposals for defining finite "random" sequences (and Knuth discusses these, also), there is a fundamental problem. The problem is that a truly "random" sequence should be able to pass any and all tests for randomness that might be proposed. A finite sequence of length k contains only k degrees of freedom. Therefore, regardless of whatever other properties it might have, it could pass at most k independent tests for randomness; of necessity any further tests would find dependencies within the set. The situation is similar to the one where calculating statistics from a set of data "uses up" the degrees of freedom in the data. Indeed, the result of performing randomness tests is the calculation of statistics, and these statistics restrict the possible values of the data as surely as the simpler ones (e.g., the mean, S.D.) do. (Review our discussion about "degrees of freedom" in Chapter 4.)

Similarly, a truly random infinite sequence contains within itself all possible finite sequences. Therefore, there is no *a priori* way to distinguish, for example, between the sequence 1,2,3,4,5,6,7,8,9 and the sequence 6,2,4,1,7,8,9,5,3. Even though the second one seems intuitively "more random" to us, not only will both of them occur in a random infinite sequence, but also both of them will occur an infinite number of times!

What is the poor statistician to do?

Even though a suitable definition exists, it applies only to infinitely long sequences and clearly, we can never obtain an infinite sequence when dealing with real measurements. Does this mean that we can never consider anything random?

In practice, there are a number of things we can do. Basically, they all come down to considering any finite set of results to be a sample from an infinite population. Just as any random sample represents the characteristics of the population, so does this sample. Furthermore, when the characteristics of the population are known, the expected characteristics of samples can be calculated. Such calculations will, of course, include the distributional requirements, and confidence limits corresponding to the various confidence intervals of the distribution.

Then it is a matter of specifying which characteristics are appropriate to perform hypothesis tests against. Since the properties of the hypothesized population are known, the properties of samples are known. We can then turn the argument around and say that if we test the sample and it passes our tests, then it is safe to consider that the population from which the data were derived was random.

The only question, then, is what tests are suitable ones to apply, considering that, as we have noted before, for a set of k numbers no more than k tests can be applied. In general, usually only a few tests are applied in any one case. In some cases, very careful workers dealing with unusually sensitive situations use a suite of tests that check characteristics of particular concern to their study; however, such scientists already know what they are doing and do not need our advice.

Empirical testing of this sort is, in the end, about all that can be done with actual data once they have been collected. The randomness of a set of numbers is only a function of how well they adhere to whichever set of properties of random numbers we choose to test. Knuth recognizes the importance of this: his chapter on random numbers includes fully 76 pages on testing sequences for randomness—as compared with only 37 pages dealing with how to generate, using a computer, number sequences that behave randomly and, as we have seen above, 30 pages dealing with the meaning of "random". As we noted before, Knuth's interest is in testing the output of a computer algorithm that produces sequences purported to be random; however, many of them are suitable tests for *any* sequence purported to be random. Knuth does, in fact, use statistical tests to check the quality of the computerized random number generators; however, his emphasis is on the computer rather than on the statistics, so that reviewing his techniques from the statistical point of view is a valuable exercise.

There are two caveats to beware of in performing such tests. The first is that it is never possible to prove that a sequence is random. The reasons for this are similar to the reasons we could not prove the purity of a jar of crystals—as we have stated many times before, statistics can disprove, but can never prove.

The second caveat deals with the number of tests to perform. It can be demonstrated that, for any given test, it is possible to generate a sequence that satisfies that test and yet is not random. A very simple example of this is the use of distribution tests, such as the one we applied in Chapter 23 to a table of numbers. In that case the test failed because the numbers were in fact, generated non-randomly. However, small variations in the table would have caused it to pass the distribution test. Suppose we took a set of such numbers, and then *sorted* them. Clearly, the distribution of the numbers would be unchanged, and they would still pass this distribution test. Equally clearly, however, they would be non-random.

Consequently, it is necessary to perform several tests and apply the basic premise of statistics: If a given sequence passes all tests, then we consider that it would have been too unlikely for that to have occurred unless, in fact, the sequence was indeed random.

At the other end, we have already seen that it is not possible for a sequence of k numbers to pass more than k independent tests. Overtesting will reveal that no finite sequence is random. Furthermore, since the results of statistical tests are probabilistic quantities, we must review a pertinent result from probability theory: The probability of occurrence of any one of a set of events, the individual probability of each being P_i, is given by the formula:

$$P = 1 - (1 - P_i)^n \tag{24-1}$$

Consequently, if two tests of a set of random numbers are performed, each at $\alpha = 0.05$, then α for the joint test is $1 - 0.95^2 = 0.0975$. When more tests are performed, the probability of at least one of them producing a statistically significant result by chance alone increases continually. To circumvent this, a two-pronged approach is needed. First, the α-level at which the individual statistical tests are being performed should be adjusted to compensate for the change in the effective α-level of the multiple tests. Secondly, a minimum suite of tests should be used, to guard against gross departures from randomness.

So let us examine some tests that are appropriate for testing the randomness of a sequence. The basic concept of one large group of tests is to describe a characteristic that random numbers should have, calculate the theoretical distribution of the numbers in terms of the categories of that characteristic, and finally, to perform a χ^2 test of the actual distribution against the theoretical one.

Knuth presents a number of tests. Some of them are more appropriate as tests of computer-generated random numbers than of the randomness of real data; one such test is the cycle-length test. Computer-generated random numbers always have some cycle because there are only a finite number of different values that a computer can hold (generally 2^n, where n is the length of the computer word) and since the numbers are produced algorithmically, any given random number is always followed by the same one, a property that real random numbers will not have.

Some of the tests have interesting names (e.g., the "poker" test, the "coupon collector's" test); however, there are too many for us to examine all of them here. Nevertheless, we will look a few of the more interesting ones.

The first one of interest is the one that we have already seen in Chapter 23, the distribution test. This test is particularly interesting because it can do much more than test for randomness. In fact, with real data, this test is more useful when it is used to verify that the errors (or the data, as appropriate) follow some assumed distribution, rather than as a test of randomness *per se*.

The most common assumption about errors is that they are Normally distributed. Thus, for a given set of data, the values can be categorized according to their values, then compared with a list of expected values and χ^2 computed. Using an appropriate list of expected values is the key to testing the distribution.

For example, in Chapter 6 we presented data representing a 100% line of a spectrophotometer. Are those values Normally distributed?

To perform this test, we first must compute the number of standard deviations that each point is from the mean, in order to be able to compare it to the distribution table for the Normal distribution. Thus, we compute the mean of the data (0.9980031633) and subtract it from each datum, then use the standard deviation of all the data (0.03663813068) to divide each datum. The resulting numbers, shown in Table 24-1, have a mean of zero and a standard deviation of unity.

The next step is to break the distribution up into ranges, to count the number of data points in each range, and to compute χ^2 for the comparison of those counts with the theoretical value obtained from the Normal distribution. The choice of the number of ranges and the ranges to use is not fixed by hard and fast rules. The chief rule that governs the choice is the one that states that the expected value for the count in each range (which is equal to np) must be at least 5. For this example, we have chosen the ranges for the standard deviation to be: below -1.5, -1.5 to -0.5, -0.5 to 0, 0 to 0.5, 0.5 to 1.5, and above 1.5. This set was picked in order to demonstrate that the ranges need not all be of the same size, nor need they all have the same expected value for the count. Other sets of ranges are certainly

TABLE 24-1

Data from 100% line converted to the number of standard deviations from the mean.

0.5117	0.1022	−0.5419	0.1050	1.0906	1.4730	0.7994	−0.6000
−0.5173	1.0098	1.3834	1.2481	1.4528	0.5076	−0.8809	−1.1238
0.1413	1.3335	0.7756	0.2128	0.7947	1.4506	1.2674	0.3487
0.3910	0.9702	−0.0642	−0.8462	−0.1887	0.1609	−0.1275	−0.2569
−0.3312	−0.6022	−0.4439	0.9888	1.5592	0.0804	−0.4769	0.7489
1.5281	1.0046	0.3231	0.6187	1.5488	1.2467	0.4006	1.0237
1.8398	0.7328	−0.7982	−0.6778	0.2592	0.1075	−0.3033	−0.1205
−0.4619	−0.9123	−1.0345	−1.2338	−1.1415	−1.4390	−1.6574	−0.7523
−0.4076	−0.5613	−0.3778	−0.7987	−0.9941	−0.4398	−0.6093	−1.4393
−1.489	−0.4900	0.7538	1.8144	1.9850	1.0008	0.2674	−0.6658
−1.9298	−2.1328	−1.9811	−1.5722	−0.8467	−1.0766	−1.3929	−1.1743
−0.2452	1.4279	1.2377	−0.4944	−1.4098	−0.9114	0.6935	0.8861
−0.0410	−0.3882						

TABLE 24-2A
Count of points in Table 24-1 under the breakpoints used, compared with the theoretical count, based on the cumulative Normal probability distribution.

Breakpoint count	Actual probability	Cumulative probability	Theoretical count
−1.5	5	0.0688	6
−0.5	33	0.3075	30
0	52	0.5	49
0.5	64	0.6915	68
1.5	92	0.9322	92

possible, and some might even be preferable. For example, selecting the ranges so that the expected count in each would be exactly five would maximize the number of degrees of freedom for the χ^2 test. (Exercise for the reader: redo this entire test using that set of ranges for the comparisons. As another exercise, redo the test using ranges of equal width.)

The theoretical values for the corresponding probabilities are obtained from the table of the cumulative normal distribution. The details of this, as well as the data for the actual values are shown in Table 24-2A. The actual counts are most easily obtained by simply counting the number of values that fall within each range. The theoretical values are obtained by calculating the differences between the cumulative probabilities corresponding to each end of the range.

For completeness, in Table 24-2A, we show the cumulative counts for the actual data below each breakpoint, along with the cumulative Normal probability distribution for the same breakpoints. Since there are 98 data points, $n = 98$ and the expected (i.e., theoretical) values for the counts (i.e., np) are then equal to the probabilities times 98; these values are also shown in Table 24-2B (rounded to

TABLE 24-2B
Calculation of χ^2 for these data.

Range	Actual count	Expected count	$(o - e)$	$(o - e)^2/e$
Under −1.5	5	6	1	0.166
−1.5 to −0.5	28	24	4	0.666
−0.5 to 0	19	19	0	0
0 to 0.5	12	19	7	2.579
0.5 to 1.5	28	24	4	0.666
Above 1.5	6	6	0	0
				Sum = 4.077

the nearest integer). From these values, it is trivial to compute the values in each range, which we will use for the computation of χ^2. Note that the cumulative values in Table 24-2A cannot be themselves used directly to perform the χ^2 test (well, they could, but that would be a misuse); since the cumulative counts at the higher breakpoints include data already used for the lower ones, these numbers are not independent.

The details of the χ^2 test are shown in Table 24-2B. Since this is similar to the ones we have done before, we will not dwell on it, but simply note that the 95% confidence interval for six degrees of freedom is 1.24 (= 14.45); therefore, there is no evidence of non-normality. Note the double negative. This form of the statement is necessary since we have not proved the distribution normal, but simply failed to prove it non-normal. This is the bugaboo of statistics, but is a crucial consideration in avoiding misuse.

Another important test is the serial test; we have seen this before also, when we were discussing contingency tables. Basically, this test operates on pairs of data points and verifies that the data following the data in any range are distributed properly. The sorted data that we described above would fail this test. This test can be extended in two ways. The first way to extend it is to test data pairs separated by 1, 2,... up to k intermediate points. Truly random data should show the same distribution regardless.

The second extension is to use groups of more than two numbers. For example, for groups of three numbers, the contingency table for the two points following data in any range should not depend upon the range the first number lies in. A practical difficulty arises in applying this test, because it requires an enormous amount of data to insure that the categories corresponding to the combined smallest counts all have at least five members.

Knuth also lists a number of very simple tests that can be applied, although they may not be χ^2 distributed. One such test is the sign test: Half the data should be above the mean and half below. Since the mean error is zero, this equates to half the numbers being positive and half negative. This is a special case of the distribution test, where the number of plus (or minus) signs should follow a binomial distribution with $p = 0.5$.

Another test, simple to understand and recommended as being a fairly strong test of randomness, is the runs test. Basically, a "run" is a subsequence of the data to be tested in which the values are continually increasing. The first decrease in the value of a point (compared to the previous point) ends that run. If two values show successive decreases, the first value is a run of length one. In use, either the total number of runs, or the number of runs of various lengths are counted, and compared with expected values. Unfortunately, these are *not* distributed as χ^2, so special tables must be used, although the principle is the same as the previous tests. Statistical tables of the expected values for runs are available (see, for example, Reference (24-2)).

References

24-1. Knuth, D.E., *The Art of Computer Programming*, Addison-Wesley, Menlo Park, CA, 1981.

24-2. Dixon, W.J., and Massey, F.J., *Introduction to Statistical Analysis*, 4th ed., McGraw-Hill, New York, 1983.

25

THE F STATISTIC

We are back to talking about statistics that apply to continuous variables, as opposed to those we dealt with in the last few chapters, which apply to discrete quantities, i.e., where data are generated by counting something.

The next statistic we will consider is the F statistic. The F statistic, along with the Z, t, and χ^2 statistics, constitute the group of what we might think of as the basic, or fundamental statistics. That is because collectively they describe all the relationships that can exist between means and standard deviations. In Chapter 20 we presented a list summarizing the various relationships that can exist between a pair of the population and sample values of means and standard deviations, with one exception conspicuous by its absence: we did not list there the statistic that describes the distribution of the standard deviations or variances of two samples.

The reason we did not list it is because at that point we had not yet discussed that distribution, but, as you might have guessed by now, that is what we are going to discuss here. If you guessed that much, then surely you have also guessed that the F statistic is the statistic that describes the distribution of the ratios of variances of two samples. Strictly speaking, the F statistic (which follows the distribution called the F distribution) describes the following ratio:

$$F = \frac{S_1^2/\sigma_1^2}{S_2^2/\sigma_2^2} \tag{25-1}$$

This equation means that the relationship between any two sample variances whatsoever can be described by the F distribution. The fact that each sample variance is related to its own population variance means that the sample variances being used for the calculation need not come from the same population; hypothesis testing can be done using variances that are from different populations. This is an enormous departure from the assumptions inherent in all the previous statistics that we have brought under consideration.

There are many important cases where $\sigma_1 = \sigma_2$ (or only one population is involved so that σ_1 *is* σ_2). In these cases Equation 25-1 reduces to $F = S_1^2/S_2^2$. Nevertheless, the fact that Equation 25–1 represents the full expression of the calculation of the F statistic means that this statistic is a much more powerful statistic than the others we have so far considered, as well as having a broader range of application. In fact, in addition to its use as a means of comparing different sets of data, in the sense that the Z, t, etc. statistics are, the F statistic is at the heart of the statistical technique called analysis of variance, about which we will have more to say in the future.

Formally the F distribution is defined by the equation:

$$F = \frac{\Gamma\left(\frac{f_1+f_2}{2}\right)}{\Gamma\left(\frac{f_1}{2}\right)\Gamma\left(\frac{f_2}{2}\right)} f_1^{f_1/2} f_2^{f_2/2} \frac{\frac{s_1^2}{s_2^2}}{\left(f_2 + f_1\frac{s_1^2}{s_2^2}\right)^{(f_1+f_2)/2}} \tag{25-2}$$

where s_1^2 and s_2^2 represent the variances being compared, and f_1 and f_2 represent the degrees of freedom corresponding to those two variances.

The fact that it is necessary to specify the degrees of freedom corresponding to both the numerator and denominator of the ratio of the variances gives rise to an interesting way of looking at the relationship between the various statistics we have so far investigated.

The Z statistic, describing the Normal distribution, requires only one table label (the probability level) to be specified, in order to fill in the table of the Z distribution. In this sense it requires only a linear, or one-dimensional table.

The t statistic and the χ^2 statistic each require two table labels (the probability level and the degrees of freedom), along two different axes, in order to lay out the table to be filled in; these are thus two-dimensional tables.

The F statistic, as we have seen, requires three table labels: the probability level and the two degrees of freedom; the F distribution thus requires a three-dimensional table to provide fully for the values that are needed. Of course, true three-dimensional tables are effectively unknown (not counting the ones, say, that computers are placed on!) Thus, F tables, whether in statistics texts (25-1) or compendia of statistical tables (25-2) are laid out as large sets of two-dimensional tables; commonly the different tables correspond to different numbers of degrees of freedom for the denominator variance, while in each table the two axes represent the numerator degrees of freedom and the probability level. In regard to terminology, the critical value for F is written to reflect all these factors; using the formalism of hypothesis testing, the critical value of F is written as $F(1 - \alpha, f_1, f_2)$. For example, $F(0.95, 12, 17)$ refers to the critical F value for a 95% confidence interval (i.e., $\alpha = 0.05$), with 12 degrees of freedom in the numerator variance and 17 degrees of freedom in the denominator variance.

The procedure for using the F distribution when performing hypothesis tests is similar to the procedure for the other statistics. The formalism that we recommended in Chapter 18 is eminently applicable; the only difference is that, when examining the table of critical values of F, it is necessary to take extra care to ensure that the correct entry is being used, i.e., that the page for the correct degrees of freedom for the denominator variance has been found and that the correct entry for numerator degrees of freedom and probability level is used.

At this point our readers might be wondering why we have delayed introducing the F statistic until four chapters after we went to all the trouble to make a table of the characteristics of the other fundamental statistics. Well, frankly, we have been stalling. There is a reason for this, and that reason is tied into one of the fundamental reasons we are writing this book in the first place. This reason is to teach enough statistics to prevent misuse; to allow our readers to learn enough statistics to know where it can or will be useful, while also teaching enough to know when they are beyond their depth and need more expert assistance in order to ensure using the power of statistics properly.

The immediate problem that we foresee coming up when we bring up the F statistic is that this statistic is one in which there will tend to be vested interest in the results of some of the hypothesis tests that can be performed with it. It was to that end that we oriented the last four chapters toward a discussion of some of the pitfalls of careless, casual hypothesis testing without safeguards. Be it known here and now that that was only the tip of the iceberg of potential misuse.

The potential for misuse is greatly exacerbated by the nature of the instrumentation marketplace: as John Coates said so well:

"The scientific instrument marketplace is a fascinating arena where scientific reasoning meets headlong with blatant commercialism" (25-3).

In the bid for competitive advantage, scientific conservatism is subjugated (to say the least) to any perception of potential commercial utility. Unproven hypotheses are claimed to be solemn fact, and any new algorithm or method of data processing is often touted as the magic answer to any and all problems.

Along with this comes a good deal of paranoia. In our attempts to assemble data to use for examples of some of the principles we are explaining, we went to the FTIR manufacturers with a request for data representing a 100% line. One manufacturer refused outright to provide such data despite our assurances of anonymity, stating "A company can be put out of business by a 100% line." Those that provided data made sure that it represented the absolute best their instrument could do. As a result, all those 100% lines were so obviously limited by the digitization capability of the associated computer that the instrument characteristic was completely lost, and the data were therefore completely useless. The one 100% line that we have been able to use was provided by a university researcher who understood and was sympathetic to our requirements, and was therefore able to give us the type of data we needed; and this is the one we have been using all along. To say the least, having only one suitable data set is somewhat limiting.

In the face of such commercialism and paranoia, our current problem comes from the fact that the *F* statistic is, indeed, the one that can be used to compare instruments. But because of the nature of the instrument business, we can almost see all the manufacturers' marketing departments running around making uncontrolled measurements of all their competitor's instruments against their own, computing *F* values for all combinations of comparisons, reporting the ones that show their instrument as being "best" and justifying it all with, "Well, we computed the *F* value and Mark and Workman said this is the way to do it." Well, Mark and Workman did not say that this is the way to do it; *F* values are indeed the way to compare variances, but there is a lot of work that comes before any *F* values are calculated, or even before any data are collected. For example, as we stated earlier, verifying that actual distributions conform to the expected population characteristics is one critical part of this, but is still only the tip of the iceberg. Other critical pieces include collecting the *right* data, so that apples are not only compared with apples instead of oranges, but also with the proper variety of apple, as well as analyzing that data properly. This task is made all the more difficult by the need to avoid the introduction of inadvertent as well as deliberate bias, confounding, or misinterpretation. In addition, all this leaves open the question of whether a single (or even a few) performance criterion, even if "statistically significant", is the right way to tell whether something as complex as a modern spectrophotometer really is "best". What about other performance criteria? What about non-technical factors: reliability, complaint response, customer support (training, applications), etc.?

Let us face it, when you go into a shoe store to buy shoes, you will probably find about five pair that are satisfactory; the choice of one of the five is probably made just "because". For better or worse, the final decision on even major purchases such as modern instrumentation is at least occasionally made for completely non-objective reasons, such as rapport with the salesman, or, as in the case of shoes, just "because", more often than we might like to believe. In the face of all that, using bad science in the belief that it will aid "the cause" seems a little silly, especially when good science would do as well, and perhaps even better, by enhancing credibility.

At this point, you may be wondering if, in fact, it is ever possible to do a valid *F* test, in the face of such complications? The answer is yes, but as we have said before, nobody ever claimed that good science was easy (well, that is not quite true either, there are always those that will *claim* anything), but certainly we never did.

It occurred to us to include an exercise here that would avoid the controversy by using an innocuous example. Simultaneously we can illustrate how statistics can be used to treat data optimally. The example that we will use is from Chapter 14: "One- and Two-Tailed Tests" where, among other things, we discussed the concept of detection limits.

Among the points brought up in that discussion is the fact that, under certain conditions, the detection limit is the sum of the upper (one-tailed) 95% confidence limit of the distribution of results for a solution containing a known zero concentration of analyte added to the one-tailed 95% confidence limit for a solution containing a small amount of analyte. If the standard deviations of the readings in the presence and absence of analyte are equal, then this detection limit is twice that standard deviation. If they are not equal, then the two values should be added.

How will we know if, in fact, the two values for σ are equal? This is what an F test will tell us. If $\sigma_1^2 = \sigma_2^2$, then S_1^2/S_2^2 will be distributed as $F(1 - \alpha, f_1, f_2)$. If the calculated F is outside the confidence interval for that statistic, then this is evidence that $\sigma_1^2 \neq \sigma_2^2$. Therefore, calculating F and performing the hypothesis test will tell us what to do. If the calculated F is significant, then the standard deviation of the readings in the presence of analyte does not equal the standard deviation in the absence of analyte, and the two should be added.

If the calculated F is not significant, then the two values of standard deviation should be *pooled*, in order to determine the value of standard deviation to use. Why pool them? First of all, this will give a better estimate of σ than either of the values alone, and second, that estimate will have more degrees of freedom than either of the individual standard deviations.

This procedure, by the way, has been a very simple example of how the statistician approaches a problem. Instead of proceeding blindly, and manipulating data arbitrarily in order to obtain some guessed-at results, a statistician first does a hypothesis test to determine the true situation, and then knows how to manipulate the data properly (and optimally).

We would have included a numerical example here, simply to illustrate doing the F test. However, that is somewhat counter-productive. We expect that by this point our readers know how the formalism of a hypothesis test is set up. Varying it by simply using F as the test statistic, and looking up the critical values in an F table is hardly a major advance in statistical sophistication—and besides, Dixon and Massey (25-1) provide plenty of numerical examples if you feel you need them.

More useful, perhaps, is another illustration of how *not* to misuse this statistic. For this example we will use some results that have been previously published in the literature (25-4).

We had available data representing the measurement of the same set of samples by seven different analytical techniques, and from that devised algorithms to extract from the data the accuracy of each individual technique. From this set of seven values for error, one might want to perform an F test (or maybe more than one) to determine whether, for example, all the methods are equivalent (i.e., could have come from populations with the same value of σ). Would such a set of F tests be valid? (Exercises for the reader: answer these questions before continuing on since we discuss the answers below: what are the requirements for validity? In the referenced paper, were any of these requirements tested? If so, which ones? Are the requirements met or not?)

The answer to the main question is no, for a number of reasons. First, we have not verified the distribution of errors of those methods. We do not know, at this point, whether all the errors are random, or what their distributions are. We did do a test for independence, and found that not all the methods have independent errors; the errors of some of the methods are significantly correlated, and a few highly so. This violates the fundamental precept of independence required for hypothesis testing.

Secondly, we also tested for and found that statistically significant bias (one form of systematic error) existed in some of the methods. Other systematic, as opposed to random, errors might also exist. This violates another fundamental precept of hypothesis testing.

The bias in the data could be corrected, at the cost of a few degrees of freedom; quite likely other possible systematic errors could be identified and corrected. Having done so, there is a good chance that the correlations among the errors of the different methods would disappear. Assume this actually happened, could we *then* do F tests to see whether all the methods are equivalent?

The answer is still no. The problem now is to decide which methods we are going to compare. Remember, critical values for the various distributions used for performing hypothesis tests are always based on taking a single random sample. In the case of the errors of the seven methods, we would need to select randomly two of the seven methods. This would leave the other five untested, which might include those with a different population error.

Clearly, on the other hand, if there is a difference between methods, it would show up most surely by comparing the one with the greatest error to the one with the least. Now we have a dilemma: selecting the best and worst is surely a non-random method of selection, and the critical values in the F table do not apply. We are stuck, and the F test cannot be used.

In fact, there is a special test, known as Bartlett's test (which is nicely described by Hald (25-5), although he does not call it that), that converts multiple variances into a variable that is distributed as χ^2, and this allows a test of equality of multiple variances such as we are concerned with (as long as the requirements of randomness, etc., are met). Note, however, that this is not an F test.

Is there anything at all that we can do with the results of the study that would fit in with the topic of this chapter? Yes, there is one type of *F* test that can be done with the results described in Reference (25-4). If the methods are chosen for extraneous reasons, then one particular method may be compared with one other particular method; those two methods may be treated as though they were the only ones that exist, and an *F* test performed (again, assuming that we have first verified randomness and independence of the errors, etc.).

References

25-1. Dixon, W.J., and Massey, F.J., *Introduction to Statistical Analysis*, 4th ed., McGraw-Hill, New York, 1983.

25-2. Pearson, E.S., and Hartley, H.O., *Biometrika Tables for Statisticians*, Biometrika Trust, University College, London, 1976.

25-3. Coates, J., Spectroscopy, 4: 1, (1989), 15–17.

25-4. Mark, H., Norris, K., and Williams, P.C., Analytical Chemistry, 61: 5, (1989), 398–403.

25-5. Hald, A., *Statistical Theory with Engineering Applications*, Wiley, New York, 1952.

26

Precision and Accuracy: Introduction to Analysis of Variance

Chemists have long been concerned with the accuracy of quantitative analytical methods. The Fundamental Reviews issue of *Analytical Chemistry* has for many years periodically contained a review of statistical methods (more recently chemometric methods) used for chemical analysis. Those reviews invariably included references to papers dealing with the question of precision and accuracy and the meaning and definition of those terms. Indeed, in the early 1950s there was apparently some controversy about the meaning of these terms, and along with the controversy, a spate of papers dealing with the subject (see, for example, References (26-1,26-2), selected somewhat arbitrarily but illustrative).

At the time there was apparently a consensus reached that all error could be represented by the two terms "precision" and "accuracy", and these two terms were considered essentially equivalent to the terms "random error" and "systematic error", respectively; furthermore, the only type of systematic error considered to exist was "bias". Bias was a constant difference between the measured value of an analyte and the "true" value. In this sense of the term, it is close to our usage of the term "biased estimator" applied to a mean, although the earlier meaning implies that the amount of difference is known.

This breakdown of errors was not unreasonable at the time. With the availability of only one method of performing an analysis, the most common situation was that a given analyst would provide data that were consistently high (or low, as the case might be) except for some random variation, while a different analyst will be still different, but also consistently.

Today the situation is considerably more complicated. There is recognition that even with wet chemical methods, there are differences among laboratories as well as analysts, among analyses performed on different days of the week, with different lots of chemicals, etc. When instrumental methods are used, there are, in addition, systematic differences that can be classified as sensitivity changes, curvature (non-linearity) of response, and more subtle consistent errors that can occur with multivariate methods of data treatment. Statisticians have risen to the occasion, and have devised methodologies for dealing with such a plethora of error sources. With wet chemical methods alone, there are interlaboratory studies, "round robin" tests, and other means of identifying various error sources.

Even in the simpler case there is a problem in specifying the accuracy. This is because the usual statement of "accuracy" compares the result obtained with "truth". We have noted earlier that "truth" is usually unknown, making this comparison difficult. Furthermore, having measured a sample several times (the generally accepted way of determining "precision"), the mean value of the measurements will not normally equal μ even if μ is known, simply because of the random variability of the data. How shall we deal with this?

The statistical technique we will use is analysis of variance. Analysis of variance actually consists of a whole slew of variations of related techniques, of which we will consider only the simplest variation in this chapter. This technique will allow us to specify a set of measurements to be made and a means of analyzing those measurements that will allow us not only to unambiguously define the two terms "precision" and "accuracy" (of course, not everyone will necessarily agree with our definitions), but also in addition compute values for them, and furthermore allow us to tell if the "accuracy" is worse than the "precision", or can be accounted for solely by it.

The basis of this analysis of variance (abbreviated as ANOVA), as it is the basis of all the variations on the theme of ANOVA, is the partitioning of the sums of squares. We have introduced this concept in Chapter 7 and highly recommend that you review that chapter at this point. The differences among the various types of ANOVA result mainly from the nature of the experimental design used (i.e., which data are taken), which in turn determines which sums of squares can be calculated.

Analysis of variance calculations were first utilized in agricultural work to determine the effects of various fertilizer treatments or other growth conditions on crop test plots. The use of ANOVA expanded to animal feeding trials to compare the various effects of different diet regimes within small subsets of farm animal populations. In biology and medicine, the use of ANOVA has become popular in the determination of the effectiveness of different drugs on human or animal test populations, and of an antibiotic on suspensions of pathogenic bacteria. These are but a few of the problems that can be resolved by this powerful statistical technique, which asks the question: "Do all my sample groups have the same mean?", or restated: "Do all my different sample groups belong to the same population?" If they do not, then evidence exists that the fertilizer (or feed, or antibiotic,...) makes a difference.

Of course, the comparison of the means of two samples is accomplished by use of the Student's *t* test. But for reasons similar to the reasons we could not perform multiple *F* tests to compare more than two variances in Chapter 25, we cannot perform multiple *t* tests to compare more than two means; comparison of three or more means must be accomplished using the more powerful ANOVA. The null hypothesis in the comparison of two means is conventionally stated as Ho: $\mu_1 = \mu_2$. The alternate hypothesis is Ha: $\mu_1 \neq \mu_2$. In the use of ANOVA to compare the means of three or more populations, we define the null hypothesis as Ho: $\mu_1 = \mu_2 = \cdots = \mu_n$, and the alternate hypothesis as Ha: at least two of the means are not equal.

So let us start with a simple example of ANOVA to illustrate the power of this statistic. We bet you thought we were going to compare the lifetime of three brands of light bulbs, or maybe even compare the average test scores from several junior high schools, or another of the examples commonly used in statistics textbooks. More chemically minded folks may have guessed we were going to compare replicate measurements on a single sample from several laboratories using the same analytical procedure; or different technicians analyzing a given sample, all using an identical procedure. Sorry! Although you can use ANOVA to determine these tests, we will use a quite different example. A more detailed look at ANOVA can be found in the excellent general discussion of linear hypothesis by Julian G. Stanley from which this discussion was partly derived (26-3).

In our one-factor ANOVA, which is the simplest one to use, we are interested in determining the photometric accuracy of spectrophotometers; we will use data from three different brands of spectrophotometers for this purpose. Photometric accuracy is defined as the "closeness" in measurement of percent transmittance, absorbance, or reflectance of standard reference materials. Thus, if a 30% T neutral density reference standard were measured on an instrument with perfect absolute photometric accuracy, it would measure as 30.0000...%T, with an accuracy equal to the recording digital precision of the instrument.

To begin, we would now sample several stock (off the shelf) instruments from the facility of each manufacturer. After unpacking the crates and being assured by each manufacturer's service representatives that each instrument is within its guaranteed specifications, we would proceed to begin work.

We designate our instrument brands as A, B, and C, and we designate our individual instruments as $A_1, A_2, \ldots, A_n$ for each brand. For mathematical purposes, it is more convenient to use numbers. Thus, the notation X_{11} represents the photometric data of instrument brand A, number 1. We know there is a hypothetical value of A_1p or an expected mean value (μ_b) for each brand of instrument, but we would not expect the photometric values within each instrument type to be exactly the same. Obviously, the instruments would read differently at different times and under different temperature, humidity, and age-of-use conditions, etc. And of course the instruments will read differently due to inherent component differences even though design and manufacturing are engineered as closely as possible. On top of all this are the fundamental limitations of thermal noise and other incorrigible noise sources.

The error expected in any specific measurement would follow a notation similar to:

$$\bar{X}_{nb} = \mu_b \pm \text{error}_{nb} \tag{26-1}$$

or, the actual measured photometric accuracy for a particular measurement X_{nb} is equal to the expected mean photometric accuracy μ_b plus some positive or negative error in the measurement (denoted as error_{nb}. The symbol n represents the individual instrument number and b is the instrument brand ($b =$ A, B, or C).

Now we go to the assumption department, which tells us so far we are OK as long as we can assume that errors from all instruments are independent, that they are all normally distributed, and that they have the same variance (σ^2). For the purpose of illustrating the ANOVA technique, we are going to simply accept these assumptions, with no justification at all, despite our extensive railing against such carrying-on. In a real case, these assumptions would all have to be checked before proceeding with the mathematical analysis. The use of these assumptions tells us that because the differences between individual instruments' performance of a particular brand are all taken into consideration in the error term, or more precisely the "random error" term, the expectation of the actual photometric accuracy for each instrument measurement within a brand would be expressed as μ_b. Because all the instruments of a brand should be expected to have identical means, independent of the sample number, we can rewrite the equation for the expected mean photometric accuracy as:

$$\bar{X}_{nb} = \mu + (\mu_b - \mu) + \text{error}_{nb} \tag{26-2}$$

where $\mu = \sum \mu_b / n$, which is the average of all the measurements of all the instruments within a brand. If we set $\mu_b - \mu = f$, we can then write the equation as:

$$\bar{X}_{nb} = \mu + f_b + \text{error}_{nb} \tag{26-3}$$

In this notation f represents the factor, or effect, caused by instrument brand on the photometric accuracy measurements. At this point what is unknown is the true mean, or the expected photometric measurement value (i.e., of whatever we are using as the physical sample in order to make the measurement), the effect of brand on the measurement, and the variance of measurement values within a brand.

So with that minimum of background, let us proceed to devise our experiment. We define our population of instruments by using a random number table or some other means of randomly selecting individual sample units. With each brand of instrument, we proceed to measure the same standard reference material under the identical conditions for all brands. Our specific concern is the difference between average measurements of the three brands, namely, $\bar{X}_A$ versus $\bar{X}_B$, $\bar{X}_B$ versus $\bar{X}_C$, and $\bar{X}_A$ versus $\bar{X}_C$.

We would like to know how the variability *between* the groups compares to the variability *within* each group. We use $\bar{X}_b$ to indicate that brand's estimate of the true mean μ_b for the bth brand of instrument. We attempt to determine, then, whether any of the three brands' spectrophotometers' average photometric value differs significantly from the others. The object is to determine if all instruments are equivalent or if there are inter-instrument differences beyond what the intra-instrument factors (e.g., noise) can account for. Since the true mean (μ) is unknown, this is the only way to compare the instruments and find out if, in fact, the variability between instruments is greater than is warranted by the within-instrument variability or not.

To work through the ANOVA problem, let us assume we have 10 instruments of each brand selected at random from many production runs, and we use a 10.000% transmittance standard reference filter at a specified wavelength. Our best estimate for the value of μ_b is thus 10.000.

Table 26-1 presents the data for this hypothetical set of instruments. From our tabled data, we can see that our estimates of μ_b are equal to the averages of each column, namely: $\bar{X}_A = 9.975$, $\bar{X}_B = 9.930$, and $\bar{X}_C = 10.092$. The best estimate of μ (or the overall mean of the average of all the values) is $\bar{\bar{X}} = 9.999$. If we want to estimate the effect of brand on our average deviation (which we denoted by f_b in Equation 26-3, where the subscript b represents brand A, B, or C), we can calculate the deviation for each brand as $f_A = \bar{X}_A - \bar{\bar{X}} = -0.024$, $f_B = \bar{X}_B - \bar{\bar{X}} = -0.069$, and $f_C = \bar{X}_C - \bar{\bar{X}} = 0.093$. We also point out again here, that $f_A + f_B + f_C = 0$. We have pointed this fact out before (have you proven it since we presented it as a reader exercise in Chapter 9?), but it is important, and we will bring it up again as an important point in the use of ANOVA. From here we compare two specific quantities calculated from the original data. The first quantity is for the within-sample sum of squares, which measures the dispersion of data within a single brand. This sum of squares within brands is calculated as

TABLE 26-1
Data for the ANOVA exercise.

Instrument	X_A	X_B	X_C
1	9.945	9.996	9.888
2	9.962	9.991	9.967
3	9.957	10.005	10.071
4	10.016	9.417	10.000
5	10.020	10.632	10.005
6	10.000	9.989	9.992
7	10.011	9.061	10.172
8	9.850	9.975	9.905
9	9.990	10.070	10.311
10	10.002	10.164	10.611
Mean	9.975	9.930	10.092
Variance	0.00262	0.178	0.0490

the sum of the dispersions of each brand:

$$\mathrm{SS}_{\mathrm{within}} = \sum_b \sum_n (X_{nb} - \bar{X}_b)^2 \tag{26-4}$$

For our data, the value of this quantity is 2.211.

The second quantity is the sum of squares between groups, which measures the dispersion of the different brands' data:

$$\mathrm{SS}_{\mathrm{between}} = N \sum_b (\bar{X}_b - \bar{\bar{X}})^2 \tag{26-5}$$

For our data, this value is equal to $10 \times 0.14 = 0.140$. (Note the N factor, further described below in Equation 26-7.

Now we have

$$X_{nb} - \bar{\bar{X}} = (X_{nb} - \bar{X}_b) + (\bar{X}_b - \bar{\bar{X}}) \tag{26-6}$$

and by squaring and summing over all the readings we partition the sum of squares of each brand:

$$\begin{aligned} \sum_n \sum_b (X_{nb} - \bar{\bar{X}})^2 &= \sum_n \sum_b (\bar{X}_b - \bar{\bar{X}})^2 + \sum_n \sum_b (X_{nb} - \bar{X}_b)^2 \\ &= n \sum_b (\bar{X}_b - \bar{\bar{X}})^2 + \sum_n \sum_b (X_{nb} - \bar{X}_b)^2 \end{aligned} \tag{26-7}$$

Since, for any constant a, $\sum_n a = na$, the above equation shows how the N multiplier enters into Equation 26-4, and we have again partitioned the sums of squares. In this case, the total sum of squares ($\sum_n \sum_b (X_{nb} - \bar{\bar{X}})^2$) is partitioned into: the within-groups sum of squares ($\sum_n \sum_b (X_{nb} - \bar{X}_b)^2$) and the between-groups sum of squares ($n \sum_b (\bar{X}_b - \bar{\bar{X}})^2$). (Exercise for the reader: fill in the missing details of the partitioning process; use the partitioning in Chapter 9 as a model.)

The underlying principle is that the within-brand sum of squares when divided by its degrees of freedom is an unbiased estimator of the variance (σ^2). In our example where we have $n = 10$ instruments of each brand, we note that the error degrees of freedom $= (3 \times 10) - 3 = 27$. This number of degrees of freedom represents the total number of measurements (observations) minus the number of means (μ_b) calculated, which was 3. In contrast, the between-brand sum of squares, divided by *its* degrees of freedom ($3 - 1 = 2$) is an unbiased estimate of σ^2 only if the μ_b, the means for each brand, are all equal. This is the key point: if, in fact, the means of all the different instruments are equal, then we know *a priori* how the *measured* values of the means should behave; they should have a standard deviation equal to $\sigma/n^{1/2}$, or the variance of those means should equal σ^2/n, and this will be reflected in the between-groups sum of squares.

If the means for each brand are not equal, then this calculated σ^2 will be larger than the value calculated from true variance σ^2/n. The between-brands and the within-brand variations are independent in the statistical sense; this not only allowed the partitioning of the sums of squares, but also permits us to treat the two independently.

If we now compute the ratio of the two sums of squares due to the different sources of variation (called within and between), each divided by its corresponding degrees of freedom, we can then calculate the F statistic for the comparison of the two variances. This will tell us whether, in fact, the between-groups sum of squares is consistent with the within-groups sum of squares or if they contain more variance than can be accounted for by the within-groups sum of squares. For our data

$$F = \frac{0.140/2}{2.211/27} = 0.855$$

If we establish our hypothesis testing, we would formulate a null hypothesis as well as an alternative hypothesis. The null hypothesis would indicate that there are *no* differences between the μ_b (the mean measurement values for each instrument brand). The alternative hypothesis would indicate that there *are* differences between the means of each brand of instrument. The calculation of the F statistic allows us the use of a statistical parameter to compare variances. The F statistic follows the F distribution, where the numerator and denominator degrees of freedom are designated and where α is used to specify the level of confidence by which to make hypothesis tests. An F test at any level of significance α causes us to reject the null hypothesis when the F statistic is too large, and to accept the alternative hypothesis. When the F test is smaller than the tabled F value at the specified degrees of freedom and level of significance, then the null hypothesis is accepted.

So in the analysis of variance method, we have calculated two estimates for the variance of the individual populations using two computational methods. If there is no effect due to the use of the different populations (i.e., brands), then the two estimates should agree. If the two methods of estimating the variances result in answers which are too far (compared to the level of significance) apart, then the assumption that all the sample means are the same is shown to be wrong, and we reject the null hypothesis. The F distribution allows us to determine what "too far" means in an objective numerical sense. The F statistic is also termed the F ratio and it allows us to measure whether our two estimates of variance around a mean value are from the same population (are alike) or are from different populations (not alike).

Now, if we wish, we can relate the two measures of the variance to the terms "precision" and accuracy". The square root of the within-group variance can be easily seen to correspond to what all chemists would agree represents "precision", once we allow the assumption that all brands of instruments have the same value of within-brand variance. At this point, we reiterate that this was an assumption, required to proceed with the analysis. If the assumption fails, then two things happen: first, we cannot define precision for the analysis, because it becomes dependent upon the brand of instrument used. Second, we must find a referee for the battles between instrument manufacturers.

The accuracy of the analytical procedure is closely related to the between-groups variance. If the F value of the test turns out to be non-significant, then the accuracy is the same as the precision. If the F value is significant, then the accuracy can be derived (or defined, perhaps) in terms of the between-groups variance, but there are some subtleties involved, which we will defer, noting that a modification of the ANOVA concept, called components-of-variance analysis, is needed.

Note that, while this procedure allows quantification of overall precision and accuracy of the spectrophotometric method, it does not compare the different instruments. This is just as well, since as we discussed before, that is a more complicated topic than these simple statistics will deal with, and would require considerably more complicated ANOVAs, with correspondingly more complicated analysis and interpretation.

In our case, again:

$$F = \frac{\text{between-brand variance estimates}}{\text{within-brand variance estimates}}$$

TABLE 26-2

General ANOVA table [for one-factor test with b treatments and N total observations (measurements)].

Source of variation	Degrees of freedom	Sum of squares	Mean square	F ratio	Critical F ratio
Mean	1	$N \times \bar{\bar{X}}^2$	MSB = SSB/$(b-1)$	$\frac{\text{MSB}}{\text{MSW}}$	***
Between treatments****	$b-1$	SSB*	MSW = SSW/$(N-b)$		
Within treatment	$N-b$	SSW**			
Total	N	Sum of squares of all data			

*$\text{SSB} = \sum_n \sum_b (\bar{X}_b - \bar{\bar{X}})^2 = N \sum_b (\bar{X}_b - \bar{\bar{X}})^2$.

** $\text{SSW} = \sum_n \sum_b (X_{nb} - \bar{X}_b)^2$, where $\bar{\bar{X}}$ is the grand mean of all the data.

*** The critical F ratio is determined from tabled values of the fractional values of the F distribution, for given values of numerator and denominator degrees of freedom.

**** The term "treatment" in this context is a piece of statistical jargon that refers to any physical consideration that might cause a difference between sets of data. In the exercise we present, each different manufacturer represents a different "treatment" of the spectrometers.

To calculate the within-brand variance estimate, we use the pooled value, which we calculate as

$$\text{mean square error (MSE)} = \frac{\sigma_A^2 + \sigma_B^2 + \sigma_C^2}{3}$$

where the various σ_b^2 indicate the variances of each brand.

To calculate the between-brand variance estimate, we calculate the variance of the set of the three brand sample means as

$$\frac{(\bar{X}_A - \bar{\bar{X}})^2 + (\bar{X}_B - \bar{\bar{X}})^2 + (\bar{X}_C - \bar{\bar{X}})^2}{2}$$

The 2 in the denominator comes from the $n - 1$ term, where n is now 3, the number of different brands. Now the ratio of the two estimates of variance is our F statistic, as shown above.

Under the null hypothesis, the numerator and denominator are actually two estimates of the same variance—namely, the population's common or overall variance. We would expect that if both estimates of variance were equal, the F statistic, or F ratio would equal 1.0. To simplify and summarize any analyses of variance application, one often constructs an ANOVA table. The general structure of the ANOVA table, for the exercise we have performed here, as well as the various computational formulas are shown in Table 26-2.

Exercise for the reader: interpret the results of the ANOVA. We will discuss the matter in greater depth in the next chapter. For now, we will continue to admonish against inappropriate use of these concepts. For example, we have just finished what appears to be an analysis of a situation that seems to result in estimates of the accuracy of an instrument. In fact, it is no such thing. There is a large piece missing from what a careful analysis would require, and that piece is a demonstration that, in fact, all instruments of the same type show the same behavior. Without that missing piece, what we have done here is nothing more than a pedagogic exercise to demonstrate the method of analysis, and says nothing about the performance of the instruments.

We recommend Reference (26-4) as a great help to beginners. It is a broad-based statistical primer, and pages 213–237 contain the ANOVA chapter.

References

26-1. Wernimont, G., *Analytical Chemistry*, 23: 11, (1951), 1572–1576.

26-2. Dyroff, G.V., Hansen, J., and Hodgkins, R., Analytical Chemistry, 25: 12, (1953), 1898–1905.

26-3. Kruskal, W., and Tanur, J., *International Encyclopedia of Statistics*, The Free Press, Division of MacMillan Publishing Co., Inc., New York, 1978.

26-4. Finkelstein, M., and McCarty, G., *Calculate Basic Statistics*, Educalc, Laguna Beach, CA, 1982.

27

Analysis of Variance and Statistical Design of Experiments

In Chapter 26 we presented the framework for performing analysis of variance in the form of a generalized ANOVA table. Using an analysis of variance table in the proper format is a formalism, just as performing hypothesis tests using the proper format is a formalism. The formalism of the ANOVA table serves the same function, and has the same importance, as the formalism for hypothesis testing: it helps to avoid blundering into pitfalls.

In Table 27-1 we present the same table, but now it is filled in with the results obtained from the data we used in Chapter 26. While the interest is still the same, i.e., to determine whether the factors (different manufacturers in this case) were all the same, the formalism of the ANOVA table specifies that the computations be performed in the order indicated. (Exercise for the reader: to verify your understanding, redo the computations and check that you get the same answers.) Thus, what we are interested in is whether the differences between the data from the different instruments are greater than can be accounted for by chance alone. We already know that by chance alone, we can expect a variation of the means that will be approximately equal to $\sigma/\sqrt{n}$, and therefore in order to show that there is some effect due to using different manufacturer's instruments, there would have to be a difference that is greater than this value. How much greater? Enough greater for it to be too unlikely to have arisen by chance alone. This is determined by performing a hypothesis test—now you can see where hypothesis testing becomes important.

TABLE 27-1
ANOVA table for data from Chapter 26 [one-factor test with 3 treatments and 30 total observations (measurements)].

Source of variation	Degrees of freedom	Sum of squares	Mean square	F ratio	Critical F ratio
Mean	1	2997.6204			
Between treatments	2	0.1409	0.0704	0.915	$F(0.95,2,27) = 3.35$
Within treatment	27	2.0793	0.0770		
Total	30	2999.840			

The statistic we use for this test is the F statistic, which compares variances. Therefore, instead of computing whether S_{between} instruments (in statistical parlance "between" is often called "among") is equal to $S_{\text{within}}/n^{1/2}$, the test is actually performed in a manner which is algebraically equivalent, but somewhat more formalized, by testing $n \times S^2_{\text{between}}$ against S^2_{within} directly. The advantage of doing it this way is that it follows directly from (and therefore conforms to) the partitioning of the sums of squares. This relates the computations to the theoretical concepts in a rigorous fashion.

Since the ANOVA table is generated directly from the partitioning of the sums of squares, one column is called, appropriately enough "sums of squares", and contains the corresponding sums of squares as entries. Note how the partitioning of the sums of squares assures that the sums of squares due to all sources (including those due to the mean) add up to the total sum of squares of all the data. This is one check on the accuracy of the calculations, and of the model—if a source of variation is taken out of the data but not included in the ANOVA table, the equality will not hold.

When the sums of squares are divided by their corresponding degrees of freedom, the resulting quantities are called "mean squares", and it is these mean squares that are used in the hypothesis tests. An important point to note is that, with some exceptions, the mean squares are *not* the variances due to the corresponding factors, although there are close relations between the two quantities, and the variances can be computed from the mean squares (we will see this below). Another, although less important, point to note is that for the total sum of squares, and for the sum of squares due to the mean, the mean square is not computed. This is because neither of those quantities is part of the variance of the data, so the mean square due to them is not part of the analysis of variance.

In Table 27-1 we see that the F found from the data is less than the critical F for these data, therefore we conclude that in this case there is no differences between instruments beyond what can be accounted for by the random disturbances of μ_b by the noise level of the instruments. Since any effect that might exist is so small that it is lost in the noise, we must conclude that to the level we have tested the instruments, there are no differences between them. (Exercise for the reader: what would we have to do in order to detect possible differences that might exist?)

In Table 27-2A we present a set of very similar data (note that these data are also synthetic), and in Table 27-2B we present the ANOVA table for these data. Note that the first two columns of data are the same as the data in Chapter 26. In this case, however, the third column of data is larger than before, and the "between treatments" entries of the ANOVA table reflect this. Now the F ratio for the between-treatments differences shows that they are now indeed larger than can be accounted for solely by the within-treatments variations. (Exercise for the reader: for more practice, do the calculations for this new set of data, and reproduce this ANOVA table.)

The question that then becomes of interest is: what is the variance of the between-treatment differences? Or, to phrase it another way, what is the standard deviation of the differences between the instruments?

We have noted above that the mean squares do not, in general, represent the variances due to the factors. Let us explore this. For the one-way analysis of variance, which is the case we are discussing, the within-treatment mean square *is* the noise variance of the instruments (you will just have to take our word for that). However, the between-treatment mean square contains two contributions: the variance due to the differences between instruments, plus the contribution from the variance due to the noise. The contribution from the noise variance is small (since all 10 instruments' data from each manufacturer is averaged, in these data the noise contribution to the total between-manufacturers mean square is equal to $S^2_{\text{within}}/10 = 0.0077$), but it is non-zero; therefore, the mean square must be corrected for this quantity in order to obtain the variance (and thus the standard deviation) of the differences between instruments. Thus, for these data,

TABLE 27-2A

Synthetic data for ANOVA computations, using data with more "between" variability than the data used for Table 27-1.

9.945	9.996	10.088
9.962	9.991	10.167
9.957	10.005	10.271
10.016	9.417	10.200
10.02	10.632	10.205
10	9.989	10.192
10.011	9.061	10.372
9.85	9.975	10.105
9.99	10.070	10.511
10.002	10.164	10.811

TABLE 27-2B
The ANOVA table for the data of Table 27-2A.

Source of variation	Degrees of freedom	Sum of squares	Mean square	F ratio	Critical F ratio
Mean	1	3039.6300			
Between treatments	2	0.7788	0.3894	5.07	$F(0.95,2,27) = 3.35$
Within treatment	27	2.0793	0.0770		
Total	30	3042.479			

the inter-instrument variance is $0.3894 - 0.0077 = 0.3817$, and the standard deviation of the differences between the instruments is 0.6178.

At this point the reader might ask why the same computation cannot be done even if the hypothesis test shows no statistically significant difference between the two measures of variance. We have already mentioned that the reason is because there is no variation not already accounted for by the "within" variability and therefore there is not any left to attribute to the "between". However, this bare statement will probably not satisfy many readers, so to make the case stronger, we present more synthetic data in Table 27-3. In this case, there is still less "between" variability than was present in the original data, although the "within" variability is still the same. If we attempt to compute the standard deviation of the "between" variability of the data in Table 27-3, we arrive at a remarkable result: the calculation of the corrected variance for the "between" variability becomes: $0.000654 - 0.00770 = -0.00704$. To calculate the standard deviation of the "between" differences will now result in an attempt to take the square root of a negative number! We present this to emphasize once again that obeying the rules of statistics is not a matter of "mere statistical niceties" but is required to avoid disaster.

TABLE 27-3A
Synthetic data for ANOVA computations, using data with less "between" variability than the data used for Table 27-1.

9.945	10.026	9.768
9.962	10.021	9.847
9.957	10.035	9.951
10.016	9.447	9.880
10.020	10.662	9.885
10.000	10.019	9.872
10.011	9.091	10.052
9.850	10.005	9.785
9.990	10.100	10.191
10.002	10.194	10.491

TABLE 27-3B
The ANOVA table for the data of Table 27-3A.

Source of variation	Degrees of freedom	Sum of squares	Mean square	*F* ratio	Critical *F* ratio
Mean	1	2981.4288			
Between treatments	2	0.001308	0.000654	0.085	*F*(0.95,2,27) = 3.35
Within treatment	27	2.0793	0.0770		
Total	30		2983.501		

Therefore, it is never correct to try to calculate the standard deviation of any source of variance unless an *F* test has been done first, and that *F* test has shown that there is, indeed, some variance due to the source for which you want to calculate.

All the discussion to this point was based upon the simplest of the possible "statistical experimental designs" (as they are called). The term "statistical experimental design" covers a lot of territory and, indeed, there are many designs for experiments that have nothing to do with examining or determining the sources of variation in an experimental setup. However, there is a large and very important class of designs for experiments that have that very purpose as their main goal. It is important to note here that the type of experimental design used goes hand-in-hand with the way the data obtained from that design can be analyzed, and also determines what information can be gleaned from the experiment.

In the example we have been using, there is only one factor, the manufacturers of the instruments under test. However, even in this case there are characteristics of the factor that are important to the nature of the experiment and to the analysis of the data. These characteristics are:

(1) Whether the level of the factor is reproducible [i.e., if an instrument from any given manufacturer is used, then (at least presumably) it will always have the same systematic effect on the mean response as any other instrument from that manufacturer]. Some factors are not reproducible, because they are inherently the result of some random process. Sampling error is a good example: two samples (in the chemical sense, rather than the statistical sense) will not be identical, and if this source of variation is important, it should be included in the experimental design.

(2) Whether the level of the factor is uncontrollable. (As used here, "controllable" means that the factor levels can be placed in a meaningful order; e.g., the effect due to any manufacturer's instruments is a random variable, whose relationship to the other manufacturers instruments is unknown

beforehand. For example, the effect due to the use of instruments from manufacturer #2 is most definitely *not* twice the effect due to those from manufacturer #1.)

These are two of the important characteristics that factors can have, and they have profound effects on the nature of the designs that must be used. The other things that affect the nature of the experimental design are: the number of factors (we only considered a one-factor design above, but multi-factor designs are often used), how many levels each factor can attain, whether more than one measurement can be made on a given specimen (when the testing process is destructive, no more than one measurement can be made), and (extremely important) whether the different factors can be varied independently. Secondary considerations, which have practical bearing on the design, involve the difficulty and expense of varying the factors and of performing each run of the experiment.

The experimental design determines the type of information that can be extracted from the data and therefore the type of ANOVA that must be used. While there are a number of specialized types of designs that have advantages in particular cases, most designs can be based on two types of "building blocks" for experimental designs. These building blocks depend upon whether the factors involved are controllable or not, and on whether or not they have systematic and reproducible effects. The terminology used to describe the experimental designs refers to whether factors are "crossed" or "nested" in the design.

Crossed Designs

The tests for equality of means of two samples, which we discussed in Chapter 12, is actually the prototype experiment for a crossed design (in this case with one factor). In this case the only operative factor is the one that is causing the hypothesized difference between the means of the two samples. In the terminology of statistical designs, this would be a one-factor, two-level design. The two levels refer, of course, to the two values of whatever might be causing the difference between the means of the two samples.

Similarly, the experiment we have been discussing in Chapter 26 and in the first half of this chapter is a one-factor, three-level design. Again, there is only one factor because only one thing (the manufacturer) is changing between all the different instruments. That factor, however, has three different "values"; hence it is present in the experimental design at three different levels. Parenthetically, we may note that in this design the instruments are nested within the manufacturers.

The question of nesting versus crossing does not otherwise arise until there are two or more factors to be included in the experiment; then it refers to how the factors are assigned to the various experimental units and how the data are treated. We will consider these later.

In order to create a crossed design, it is necessary for all the factors to be controllable, so that they can be set to any desired level. Then it becomes possible to expose a given specimen to any desired combination of factors.

Crossed designs are very often carried out with only two levels of each factor. The reason for this is that when all factors are applied to the set of specimens, usually all possible combinations of factors are used; this value becomes large very rapidly when more than a few factors are involved, and the time, amount of work, and expense all increase proportionately. Thus, keeping the number of levels of each factor to a minimum becomes important.

When the factors are all present at only two levels, the common nomenclature used is to describe the levels of each factor as "high" or "low". When written down, they are often as not called "0" or "1". The actual values of the factors that correspond to the two levels will depend, of course, on both the nature of the factor and the goals of the experiment.

The reason such designs are called "crossed" becomes clear when the design is laid out schematically, as in Figure 27.1. As the first representation of the experiment in Figure 27.1-B shows, having each factor along a different axis "crosses" the effects. Experimental designs such as those shown in Figure 27.1-B are called "balanced", because for each level of factor 1 there are equal amounts of each level of factor 2. Balance is a way of keeping the effects of a factor independent of the effects of other factors, an important consideration in being able to extract the effect of each factor from the data. Just as the samples can be arranged in a square to represent two factors, each at two levels, three factors can be represented as being on the corners of a cube, and so forth for more factors. Designs like these are called, not surprisingly, "factorial" designs.

Crossed designs are also used with more than two levels of each factor. Figure 27.2 shows how such an experimental design would be set up. When we remember that crossed designs are used when the various levels of each factor are expected to have systematic effects (and assuming that the effects of the factors are indeed independent), then it becomes clear that computing the average across each row will result in the best possible estimate of the effect of each level of factor 2, because it minimizes the effect of factor 1 (if such an effect exists) as well as the contribution of noise. Similarly, averaging down the columns will give the best possible estimate of the effect of the levels of factor 1. While we will defer the detailed analysis of the experimental designs until Chapter 28, here we will note that the variance (or rather, the mean square) of each factor can now be computed, with minimum interference from other factors. Another point to note in Figure 27.2 is the notation: the subscript representing the axis over which the averaging was done is replaced by a dot; this indicates that the value in that position of the table does not represent a single level of the factor, but rather the average

Factor 1

Levels

Low	High

FIGURE 27.1-A. One-factor, two-level experiment.

	Factor 1	
	High	Low
Factor 2 = High	HH	HL
Factor 2 = Low	LH	LL

Alternate representation

Factor 1	Factor 2
H	H
L	H
H	L
L	L

FIGURE 27.1-B. Two-factor, two-level crossed experiment.

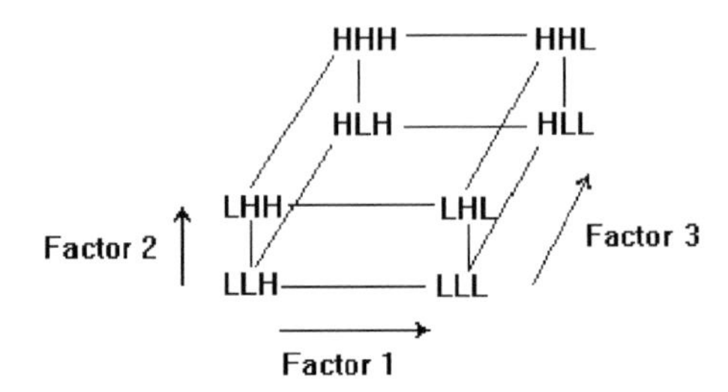

Alternate Representation

Factor 1	Factor 2	Factor 3
H	H	H
H	H	L
H	L	H
H	L	L
L	H	H
L	H	L
L	L	H
L	L	L

FIGURE 27.1-C. Three-factor, two-level crossed experiment (factorial design).

		Factor 1 Levels							
		1	2	3	4	5	...	m	
F									
A	1	nn	nn	nn	nn	nn		nn	$\bar{x}_{1.}$
C	2	nn	nn	nn	nn	nn		nn	$\bar{x}_{2.}$
T	3	nn	nn	nn	nn	nn		nn	$\bar{x}_{3.}$
O	4	nn	nn	nn	nn	nn		nn	$\bar{x}_{4.}$
R	5	nn	nn	nn	nn	nn		nn	$\bar{x}_{5.}$
2	n	nn	nn	nn	nn	nn		nn	$\bar{x}_{n.}$
		$\bar{x}_{.1}$	$\bar{x}_{.2}$	$\bar{x}_{.3}$	$\bar{x}_{.4}$	$\bar{x}_{.5}$		$\bar{x}_{.m}$	$\bar{X}_{..}$

FIGURE 27.2. A two-factor crossed design with m levels for factor 1 and n levels for factor 2.

over all levels. The grand mean of the data is marked with a dot replacing both indices; this indicates that averaging was carried out over both axes of the table.

Nested Designs

A nested design must be used when a factor is uncontrollable or the effect is non-systematic. The reason for this can be seen in Figure 27.3 where we have reproduced the generalized design for the original experiment that we used to start this whole thing off. The three column averages can be computed, as shown—no problem. However, attempting to compute the row averages is futile. Since each row represents a different instrument from each manufacturer, the inter-instrument differences are effectively the "noise", and we do not expect any systematic effect due to the noise. If the noise for

	Factor 1 Levels				
	1	2	3	4	5
N	nn	nn	nn	nn	nn
O	nn	nn	nn	nn	nn
I	nn	nn	nn	nn	nn
S	nn	nn	nn	nn	nn
E	nn	nn	nn	nn	nn
	nn	nn	nn	nn	nn
	$\bar{x}_{.1}$	$\bar{x}_{.2}$	$\bar{x}_{.3}$	$\bar{x}_{.4}$	$\bar{x}_{.5}$

FIGURE 27.3. When an uncontrolled, non-systematic variable is substituted for a factor, the analysis cannot be performed in the same manner as a true factorial experiment.

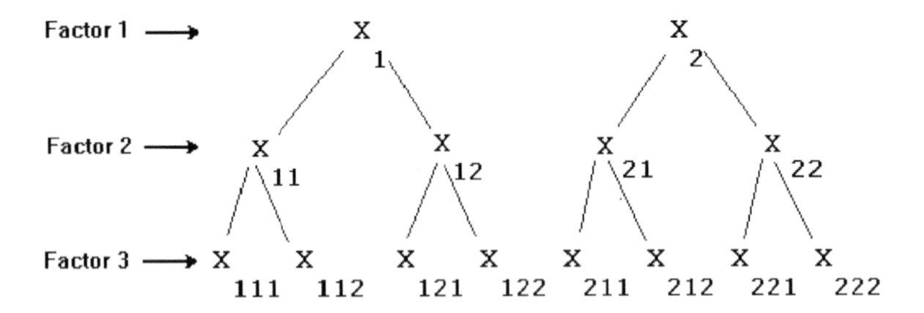

FIGURE 27.4. Variables are nested when they cannot be controlled, or the effects are not systematic.

reading #3, say, were positive for manufacturer #1, we would not expect manufacturers #2 and #3 to also be positive for reading #3. The 10 averages taken across the different manufacturers should simply have a standard deviation equal to the noise level divided by the square root of three. Thus, attempting to "cross" the noise with the true factor is a meaningless operation. (Very, very important exercise for the reader; this will test your understanding of the whole basis and purpose of hypothesis testing: suppose this experiment was done, the noise treated as though it was a factor crossed with the manufacturers (even though we have shown why this should not be done), and the F test for this "factor" turned out to be statistically significant: what would that indicate?)

In general, though, when a source of variability is uncontrollable, the factor involved must be nested within the other factors of the experiment. Figure 27.4 shows how such an experiment is set up. Note that the design allows the lower factors to be determined with more degrees of freedom (factor 1 has one d.f., factor 2 has two d.f., and factor 3 has four d.f.)

In Chapter 28 we will discuss the analysis of these designs, as well as presenting some more exotic types that have particularly useful properties.

28

CROSSED AND NESTED EXPERIMENTS

In Chapter 27 we described and discussed the two most common building blocks used for statistically designed experiments: nested factors and crossed factors. In that chapter we described how to set up the experiment but said nothing at all about what to do with the data; so here we take up the discussion from that point.

Crossed Designs

Let us first consider a crossed design, as is shown in Figure 27.2 of Chapter 27. This experiment consists of m levels of factor 1 and n levels of factor 2. Due to the way they are laid out in the data table, these are also called the "row" effects and the "column" effects, respectively. There is actually more involved here than just the nomenclature, since the use of the terms "row" and "column" imply a certain generality beyond that implied by the numbering of factors: *any* variable or type of variable can be used to indicate the difference between the entries along a row (or down a column). Measurements are made at all combinations of levels of the two factors; therefore, the data is comprised of $m \times n$ independent measurements and contains $m \times n$ degrees of freedom. Initially, assume that the effects of the two factors are independent. That is, when factor 1 has a given value, it has a given effect [e.g., when factor 1 has a value of i (that is, is present at the ith level), the effect is an increase in the measured value of one unit], and this effect is the same for all values of factor 2. The same is true, of course, for factor 2. Another way to look at this data table is as an addition table just like the one we all learned in second grade—with two exceptions: our data table does not have the factor effects increasing in nice, neat, uniform, increments; and the data table, unlike the addition table, is contaminated with noise. However, since each entry is (presumably) the result of adding the effects of the two factors, we can write it down algebraically as:

$$X_{ij} = \bar{\bar{X}} + (\bar{X}_{i\cdot} - \bar{\bar{X}}) + (\bar{X}_{\cdot j} - \bar{\bar{X}}) + e_{ij} \tag{28-1}$$

Equation 28-1 is the *model* for the experiment, and states that the value of any given datum in the table (at the i, j position) representing the experimental design is the sum of four quantities: the grand mean of all the data, the effect due to the value of the ith level of the first factor, plus the effect of the jth level of the second factor, plus the (random effect of the) noise contribution. This equality is exact, since any entry in the data table can be computed from the corresponding values of these four items.

Through the use of lots of time, effort, and paper, the sum of the squares of the data can be partitioned (guess what the exercise for the reader is!) into contributions around the grand mean of all the data (designated as $\bar{\bar{X}}$). Having done this, we can determine whether either of the factors has an effect beyond that due to random chance.

The analysis of data from an experiment such as this is done via what is called a two-way ANOVA (recall that our previous experiment—the one with the 10 instruments from each of three manufacturers used a one-way ANOVA). After subtracting the grand mean from the data, we find that the sum of squares due to factor 1 is:

$$n\sum_{i=1}^{m}(\bar{X}_{i\cdot} - \bar{\bar{X}})^2 \tag{28-2}$$

and the sum of squares due to factor 2 is:

$$m\sum_{i=1}^{n}(\bar{X}_{\cdot j} - \bar{\bar{X}})^2 \tag{28-3}$$

Now, after computing the effect of each level of factor 1 (by estimating it as the average of the data in the corresponding column) from its corresponding column, and doing the same with the rows (corresponding to factor 2), what is left is the error, and the sum of squares due to the error is:

$$\sum_{j=1}^{n}\sum_{i=1}^{m}e_{ij}^2 \tag{28-4}$$

The computations, then, are actually pretty straightforward: first the grand mean of all the data is computed and subtracted from the entire data array. Then, the mean of each column is subtracted from that column, and the mean of each row is subtracted from the row. Since the grand mean has already been subtracted from the data, the sums of the row means and of the column means are zero, as is the sum of what is left, which are called *residuals*.

The mean squares are computed (as always) as the sums of squares divided by the degrees of freedom. How many degrees of freedom for each contribution to the variance? Well, let us count them:

For the total data: $n \times m$
For the grand mean: 1
For factor 1: $m - 1$ (because the grand mean was subtracted)
For factor 2: $n - 1$ (because the grand mean was subtracted)
For the error: $nm - n - m + 1$ [Exercises for the reader (before reading the next paragraph): Why? And where does this 1 come from?]

To check our understanding, let us review how the number of degrees of freedom for the error is counted. There are nm total degrees of freedom in the data. Recall that every time a statistic is computed and used to manipulate the data, a degree of freedom is lost. In computing the factor 1 means, we compute m averages, and in computing factor 2 means, we compute n averages. The grand mean was also subtracted; therefore we lost $(n + m - 1)$ degrees of freedom due to computations on the data, leaving $nm - (n + m - 1)$ degrees of freedom for the error. We also note here that $nm - (n + m - 1)$ is algebraically equivalent to $(n - 1)(m - 1)$, an alternate form of the expressions that is somewhat more convenient for computation.

The really nice part about this type of experimental design (i.e., a crossed design with multiple levels of the factors) is that it provides an estimate of the error without having made any special effort to collect data for that purpose. In our initial foray into ANOVA, you will recall, measurements of 10 instruments from each manufacturer were needed, solely for the purpose of providing a value for the noise (or "experimental error", or whatever you want to call it).

There is an often-quoted truism in statistics, that the more *a priori* information you have about the system under study, the more information you can obtain from the experiment; however, the details of how this is accomplished are rarely described. We will provide an exception to this. We have just noted that we were able to obtain information about something not expressly included in the experiment (i.e., the noise level), without seeming to have done anything at all to allow this to happen. Where did this extra information come from? By now our reader should have guessed that there was *a priori* information, but what was it? In this two-way ANOVA the extra, *a priori* information available, is the knowledge that both factors have effects that are systematic—and that was enough to make the difference. If one of the factors was not systematic, or not controllable, then multiple data points must be collected with different values of that variable solely for the purpose of allowing an estimate of the noise level. Note that in both cases, noise is measurable—what disappeared from the experimental results was the effect of the non-controllable factor. In order to get both pieces of information, a more complex experimental design would be needed.

As in the case of the one-way ANOVA for the one-factor experiment, the formalism provides for the results to be presented in an ANOVA table. We provide the skeleton for this ANOVA table in Table 28-1. In this case, the ANOVA table is basically two one-way tables combined into one, with the common information appearing only once.

There is an item missing from Table 28-1. It is missing from the table because it was missing from our discussion. In deriving our formulas, we made the assumption that the effects of the two factors were completely independent. If that assumption does not hold, there is an extra source of variability introduced into the data. For example, consider the data in Table 28-2A. It is clear from even a casual inspection that the variability of the data in column 2 is twice that of column 1, and in column 3, three times. Analyzing it as a simple two-way ANOVA with the assumptions of independence and additivity results in the breakdown in Table 28-2B, and the ANOVA results in Table 28-2C. However, the data has a large systematic source of variability (the multiplicative factor between the variables) that is not accounted for in the ANOVA table; that is because this source of variability is not included in the model from which the partitioning of sums of squares and ultimately the ANOVA table, was derived.

Two important points: first, failing to recognize and to deal properly with systematic effects in the data is another of those ever-present pitfalls that we keep warning about. The two-way ANOVA is wrong for the data of Table 28-2; the reason it is wrong is because the residuals do not represent the random noise contribution to the variability of the data. Hence the row and column mean squares are not being compared to noise, but to another systematic variation; this is not what ANOVA is for—when only systematic variations are present, you do

TABLE 28-1
ANOVA table for a two-way analysis of variance.

Source of variation	Degrees of freedom	Sum of squares	Mean square	F ratio	Critical F ratio
Mean	1	$n \times m \times \bar{X}^2$			
Row effects	$m - 1$	SSr (see text)	MSr = SSr/d.f.	MSr/MSn	See tables
Column effects	$n - 1$	SSc (see text)	MSc = SSc/d.f.	MSc/MSn	See tables
Residuals	$(n - 1)(m - 1)$	SSn (see text)	SSn/d.f.		
Total	$n \times m$	TSS			

Notes: the lower-case r, c, and n refer to the row, column, and noise (called "residuals") contributions to the total variance. Each mean square is computed from the indicated sum of squares and the corresponding number of degrees of freedom.

TABLE 28-2A
Synthetic data showing *interaction* among the variables in a crossed design.

	Factor 1 levels		
Factor 2	1	2	3
1	1	2	3
2	2	4	6
3	3	6	9
4	4	8	12
5	5	10	15

not need statistics to analyze the situation, and, in fact, it is the wrong tool for the job (furthermore, the values for F and for the critical F are meaningless, since they are only valid for comparisons of random variables).

Secondly, the model used for this analysis is incomplete since it did not take into account all the systematic variations of the data. When the data cannot be represented solely by additive effects because a multiplicative factor is present on one or more variables, the general statistical term used is *interaction*. We will not pursue the question of interactions in ANOVA because it is, as they say, "beyond the scope of this text", but methods of detecting and dealing with it in the analysis of statistically designed experiments are available in virtually all books that discuss ANOVA (see, for example, the discussion in Reference (28-1)). However, we will pursue the question of knowing how to look for and how to recognize these systematic effects, so that you can avoid the pitfall of mistaking them for the random noise.

TABLE 28-2B
Decomposition of Table 28-2A into the contributions.

	Factor 1 levels			
Factor 2	1	2	3	Row means
		Residuals		
1	2	0	−2	2
2	1	0	−1	4
3	0	0	0	6
4	−1	0	1	8
5	−2	0	2	10
				Grand mean
Column means	3	6	9	6

TABLE 28-2C
ANOVA table for data from Table 28-2A.

Source of variation	Degrees of freedom	Sum of squares	Mean square	F ratio	Critical F ratio
Mean	1	540			
Row effects	4	120	30	12	$F(0.95,4,20) = 2.86$
Column effects	2	90	40	18	$F(0.95,2,20) = 3.49$
Residuals	8	20	2.5		
Total	15	770			

It should be clear at this point that doing calculations on data will produce numbers, but even above and beyond the haranguings we keep pouring onto your heads, it should be obvious that simply doing calculations without verifying the nature of the data will produce numbers that do not necessarily have any meaning at all, much less the meaning that is being assigned to them. The case in point is the current decomposition of the data in Table 28-2A. While we were able to create a model, partition the sums of squares and set up the ANOVA table for that data, it turned out that what we called "error" was really another systematic effect that was simply not included in the model. How could we detect this?

One way was to do exactly as we did: examine the residuals and see the pattern in them. In this case the pattern was obvious, and could be detected by direct inspection. The only thing we had to do was to actually do the looking—this is the part that often gets skipped. All the knowledge of statistics, and all the talk about performing hypothesis tests and everything else, is useless if these things are not done on a routine basis, which is why we keep railing about them.

In other cases the pattern may not be so obvious; how can these patterns be detected? In the first place, you must look for them, just as we were talking about above. To "look for" these patterns, you must first know what to look for, because you do not usually look for a particular pattern (how would you know which ones to look for?) so much as you look for any pattern at all.

Remember that the ANOVA is based on a model, of which the error term is one part. The model also specifies the properties that the error must have: it must be random, normally distributed, and have constant variance. Therefore, when performing ANOVA, one should test the residuals to see if, in fact, they have those properties. In previous chapters we discussed the question of randomness, the normal distribution, how to test for normality, and how to compare variances. Here is where those concepts are put to use: if the residuals are not randomly and normally distributed, or if the variance of the residuals is not the same at the different levels of one or more of the factors, then the model being used is not correct for the data at hand. To be sure, doing this to the ultimate degree and knowing how to correct the model is tantamount to an art, but anyone concerned with analysis of data should learn to do this in at least a workmanlike fashion.

Multivariate ANOVA

Discussion of a crossed experimental design leads more or less naturally to another topic: the ANOVA of data depending on more than one variable. The very existence of a crossed experimental design implies that two (or more) variables are affecting the data and giving rise to the values measured.

A multivariate ANOVA has some properties different than the univariate ANOVA. The simple fact of including two separate variables is enough of a clue that it is multivariate. The ANOVA for the simplest multivariate case, i.e., the partitioning of sums of squares of two variables (X and Y), goes as follows.

From the definition of variance:

$$\mathrm{Var}(X+Y) = \frac{\sum_{i=1}^{n}((X+Y) - (\overline{X+Y}))^2}{n-1} \tag{28-5}$$

expanding Equation 28-1 and noting that $(\overline{X+Y}) = \bar{X} + \bar{Y}$ results in:

$$\mathrm{Var}(X+Y) = \frac{\sum_{i=1}^{n}((X+Y)^2 - 2(X+Y)(\bar{X}+\bar{Y}) + (\bar{X}+\bar{Y})^2)}{n-1} \tag{28-6}$$

expanding still further:

$$\mathrm{Var}(X+Y) = \frac{\sum_{i=1}^{n}(X^2 + 2XY + Y^2 - 2X\bar{X} - 2Y\bar{X} - 2X\bar{Y} - 2Y\bar{Y} + \bar{X}^2 - 2\bar{X}\bar{Y} + \bar{Y}^2)}{n-1} \tag{28-7}$$

Then we rearrange the terms as follows:

$$\mathrm{Var}(X+Y) = \frac{\sum_{i=1}^{n}(X^2 - 2X\bar{X} + \bar{X}^2 + Y^2 - 2Y\bar{Y} + \bar{Y}^2 + 2XY - 2Y\bar{X} - 2X\bar{Y} + 2\bar{X}\bar{Y})}{n-1} \tag{28-8}$$

and upon collecting terms and replacing $\mathrm{Var}(X + Y)$ by its original definition, this can finally be written as:

$$\frac{\sum_{i=1}^{n}((X+Y) - (\overline{X+Y}))^2}{n-1} = \frac{\sum_{i=1}^{n}(X-\bar{X})^2}{n-1} + \frac{\sum_{i=1}^{n}(Y-\bar{Y})^2}{n-1} + 2\frac{\sum_{i=1}^{n}(X-\bar{X})(Y-\bar{Y})}{n-1} \tag{28-9}$$

The first two terms on the RHS of this expression are the variances of X and Y. The third term, the numerator of which is known as the cross-product term, is called the covariance between X and Y. We also note (almost parenthetically) here that multiplying both sides of Equation 28-9 by $n-1$ gives the corresponding sums of squares, hence Equation 28-9 essentially demonstrates the partitioning of sums of squares for the multivariate case.

The simplest way to think about the covariance is to compare the third term of Equation 28-9 with the numerator of the expression for the correlation coefficient. In fact, if we divide the last term on the RHS of Equation 28-9 by the standard deviations (the square root of the variances) of X and Y in order to scale the cross-product by the magnitudes of the X and Y variables, we obtain:

$$R = \frac{\dfrac{\sum_{i=1}^{n}(X-\bar{X})(Y-\bar{Y})}{n-1}}{\sqrt{\dfrac{\sum_{i=1}^{n}(X-\bar{X})^2}{n-1}\dfrac{\sum_{i=1}^{n}(Y-\bar{Y})^2}{n-1}}} \tag{28-10}$$

and after canceling the $n-1$'s we get exactly the expression for R, the correlation coefficient:

$$R = \frac{\sum_{i=1}^{n}(X-\bar{X})(Y-\bar{Y})}{\sqrt{\sum_{i=1}^{n}(X-\bar{X})^2\sum_{i=1}^{n}(Y-\bar{Y})^2}} \tag{28-11}$$

There are several critical facts that come out of the partitioning of sums of squares and its consequences, as shown in Equations 28-9 and 28-11. The primary one is the fact that in the multivariate case, the variances add only as long as the variables are uncorrelated, i.e., the correlation coefficient (or the covariance) is zero.

There are two (count them: two) more very critical developments that come from this partitioning of sums of squares. First, the correlation coefficient is not just an arbitrarily chosen computation, but bears a close and fundamental relationship to the whole ANOVA concept, which is itself a very fundamental operation that data are subject to. As we have seen here, all these quantities: standard deviation, correlation coefficient, and the whole process of decomposing a set of data into its component parts, are very closely related to each other, because they all represent various outcomes obtained from the fundamental process of partitioning the sums of squares. As we will see later (in Chapters 32–37), when we partition the sums of squares from a spectroscopic calibration, similar close relations hold, and we will find that many of the various statistics that are used to examine the performance and robustness of a calibration are ultimately derived from a similar partitioning of the sums of squares.

The second critical fact that comes from Equation 28-9 can be seen when you look at the chemometric cross-product matrices used for calibrations. What is the (mean-centered) cross-product matrix that is often so blithely written in matrix notation as $\mathbf{X}'\mathbf{X}$? Let us write one out (for a two-variable case like the one we are considering) and see:

$$\begin{bmatrix} \sum_{i=1}^{n} (X_1 - \bar{X}_1)^2 & \sum_{i=1}^{n} (X_1 - \bar{X}_1)(X_2 - \bar{X}_2) \\ \sum_{i=1}^{n} (X_1 - \bar{X}_1)(X_2 - \bar{X}_2) & \sum_{i=1}^{n} (X_2 - \bar{X}_2)^2 \end{bmatrix} \quad (28\text{-}12)$$

Now, if you replace X_1 by X and X_2 by Y, gosh-darn those terms look familiar, don't they (if they do not, check Equation 28-9 again!)? And note a fine point we have deliberately ignored until now: that in Equation 28-9 the cross-product term was multiplied by two. This translates into the two appearances of that term in the cross-product matrix. The cross-product matrix, which appears so often in chemometric calculations and is so casually used in chemometrics, thus has a very close and fundamental connection to what is one of the most basic operations of statistics, much though some chemometricians try to deny any connection. That relationship is that the sums of squares and cross-products in the (as per the chemometrics of Equation 28-12) cross-product matrix equals the sum of squares of the original data (as per the statistics of Equation 28-9). These relationships are not approximations, and not "within statistical variation", but are, as we have shown, mathematically (algebraically) exact.

Nested Designs

When the source of the variability is not reproducible, or the effect is uncontrollable, then it is necessary to use a nested design. As we have shown previously, nested designs are described by a diagram that looks like a tree (actually, an inverted tree). Basically, instead of setting the factors to their desired values, all the factors are kept at the same value except the one to be studied. Then the variable at the next higher level in the tree is changed and the last one changed for the new values of the other variables. We have reported such an experiment previously (28-2), and in Table 28-3 we reproduce a very small fraction the data used in that experiment. In that experiment wheat was analyzed for its protein content through the use of near-infrared reflectance analysis, and the concern was for the effect of the physical sources of variability on the results. The variables studied were: the sampling error of unground wheat, the effect of repack (including the sampling error of ground wheat), the effect of sample orientation (i.e., the rotational position of the sample cup with respect to the instrument's optics), and the noise level.

It is clear that none of these variables are controllable. It is well known that when ground wheat is measured via NIR analysis, if the sample is dumped out of the sample cup, repacked, and measured a second time, a different answer is obtained due to the changes in the surface of the sample with the new set of particles. Similarly, the effect of orientation will vary with the different packs, even though it might be possible to reproduce the position of a particular sample in its cup as long as it is undisturbed; the electronic noise is the textbook example of uncontrollable random variability.

Therefore, for this experiment, it was necessary to pack a cup, take two readings in one orientation, then two readings in another orientation, and then repeat the whole process for another pack of the sample. Thus, eight readings were required for each sample, in order to measure all three effects. Data for three samples, shown in Table 28-3, confirm that there is no relationship among the packs (four readings per pack), the orientations (two readings per orientation), or the individual readings (the individual readings have only the random instrument noise as a source of variability).

It should also be clear that it is necessary to maintain a given pack undisturbed for all four readings within that pack, and to maintain each orientation unchanged for the two readings at that orientation. What would happen otherwise? Suppose the experiment was done in such a way that for each pack of a sample, only one reading was taken (that reading would, of necessity, be at one and only one particular orientation). The differences between the readings from different packs would then include not only the pack-to-pack differences, but also the orientation differences and the variability due to the noise; these other sources of variation did not go away just because the experiment was not set up to measure them. However, the repack standard deviation would then be measured as being somewhat larger than its true value, because it includes the contribution of these

TABLE 28-3A
Data from a fully nested design.

			Sample		
Pack	Orientation	Reading	1	2	3
1	1	1	14.857	16.135	15.005
1	1	2	14.845	16.135	15.016
1	2	1	14.849	16.130	14.980
1	2	2	14.866	16.206	15.062
2	1	1	15.017	16.118	14.740
2	1	2	15.051	16.096	14.785
2	2	1	14.934	16.084	14.784
2	2	2	14.988	16.114	14.864

other sources of variation, and the different effects would not be separately determinable. The statistical parlance used to describe an experiment in which the effects of different factors are inseparable in this way is *confounded*, so when a statistician speaks of "that confounded experiment" he is not cursing the results of an experiment that did not work out the way he wanted, but merely expressing the inseparability of the effects (which may also lead to cursing of another sort, but that is not appropriate for our discussion). The inseparability of repack variation and sampling error of ground wheat, which was discussed earlier, is an example of confounded factors.

The analysis of the data is done in stages; Table 28-3B shows the sequence of steps for the data from one of the samples. First, for each pair of data points corresponding to the two readings on each orientation, we compute the difference. (Actually, what we wish to compute is:

$$\sum_{\text{samples}} \sum_{\text{readings}} (X - \bar{X})^2 \tag{28-13}$$

but as we have shown previously in Chapter 7, in the special case of two readings, this can be computed from the differences between pairs of readings.) We then compute the mean reading corresponding to each orientation, and then the differences between the two orientations for each pack. Finally, we compute the mean reading for each pack and the differences among packs. Thus, each sample then gives us four values for the differences among readings (attributable to noise) and four degrees of freedom for these readings, two differences and two degrees of freedom for orientation, and one difference and one degree of freedom for packs. We compute the sums of squares according to the formula derived in Chapter 7, and then the mean squares for each factor. (Exercise for the reader: write out the model and create the ANOVA table for the data in Table 28-3B.)

If we had taken many (ideally, infinitely many) readings at each orientation, then the average of those readings would be the "true" value of the reading at that orientation. Since we only took two readings at each orientation, the reading we obtained is perturbed by the variability of the readings; thus, the mean square of the difference between orientations contains two components of variance: the variance due to the orientation differences plus the variance of the differences between readings divided by 2. Similarly, the mean square among packs contains three components of variance: the variance among packs, plus the variance between orientations divided by 2, plus the variance between readings divided by 4. (Exercise for the reader: why 4?)

However, similarly to the case of crossed designs, an F test must show a significant increase of orientation variance over the noise before an "orientation" component of variance can be computed, otherwise similar disastrous results can occur.

TABLE 28-3B
Computations on the data from Sample 1.

Pack	Orientation	Reading	Data	Difference (noise)	Mean (orientation)	Difference (orientation)	Mean (pack)	Difference (pack)
1	1	1	14.857					
				0.012	14.851			
1	1	2	14.845					
						0.006	14.854	
1	2	1	14.849					
				−0.017	14.807			
1	2	2	14.866					
								0.143
2	1	1	15.017					
				−0.034	15.034			
2	1	2	15.051					
						0.073	14.997	
2	2	1	14.934					
				−0.054	14.961			
2	2	2	14.988					
			Sum of squares	0.002252		0.00268		0.0102
			Mean square	0.000563		0.00139		0.0102

References

28-1. Box, G.E.P., Hunter, W.G., and Hunter, J.S., *Statistics for Experimenters*, Wiley, New York, 1978.
28-2. Mark, H., *Analytical Chemistry*, 58: 13, (1986), 2814–2819.

29

Miscellaneous Considerations Regarding Analysis of Variance

We have spent the last three chapters discussing the statistical procedure called analysis of variance, but we have, in fact, only scratched the surface of this topic. There is hardly a book dealing with statistics that does not include some discussion of this topic, and there are whole books devoted exclusively to it (see, for example, Reference (29-1))—it is a major topic indeed. This is not surprising in view of the fact that analysis of variance is, as we have mentioned before, tied in very closely with what is called the statistical design of experiments. While the term "design of experiments" includes all considerations that bear on the method of obtaining useful data from physical phenomena, what is of interest to statisticians under this heading are the combinations of physical changes that are to be applied to the experimental items. The purpose of specifying particular combinations of factors is to obtain the maximum amount of information with the minimum amount of experimentation. We have already noted that whether factors are controllable or not determines whether the experimental design must be "crossed" or "nested", and we have also noted the existence of interaction and of confounding of effects. Other items of concern that may show up in the physical nature of the experiment are whether the experiments are destructive to the samples (so that it is never possible to perform exact duplicate runs), if there is drift of any sort in one or more of the factors or in the response, or if the order of performing the individual experiments might affect the results (for example, design of some instruments may cause carry-over effects from one sample to the next, so that a measurement on a given sample gives consistently higher readings if the preceding sample is high than if the preceding sample is low). All these considerations must be taken into account.

All statistical designs of experiments start with the partitioning of the sums of squares, according to the model used to describe the data. There is a somewhat subtle point here that we wish to address. The partitioning of sums of squares is an exact algebraic operation; not subject to the variability of sampling that we have been discussing for so long. So what happened to that variability, especially since we have been claiming all along that taking variability into account is the basis of the science of statistics? The point is that even though in one particular dataset the different sums of squares add up exactly, each individual sum of squares is a sample of what might be considered the universe of all possible sums of squares due to that phenomenon: the population of sums of squares, if we may use that term. Thus, another, similar, dataset will have different sums of squares corresponding to each source of variability than the first, but those sums of squares will also add with algebraic exactness, for that particular set of data. Remember, it would not be unusual for the component sums of squares to have changed because the total sum of squares will also have changed, allowing the various sums of squares to still add exactly.

The problem of minimizing the amount of experimentation required to obtain the required information is itself a very large subtopic of statistical experimental design. Often there are many factors that must be investigated in a particular experiment. This translates into a good deal of time, effort, and, most important, expense. We have noted previously that one way to minimize the number of experiments is to have each factor present at only two levels. This often fits in well with other aspects of the experiment, since the two levels can represent "before" versus "after" a particular phenomenon is allowed to act, or perhaps can represent the "treated unit" versus the "control unit". Experiments with many factors, each at two levels, are termed "factorial designs".

However, even when factorial designs are used, inclusion of many factors in the experiment leads to a requirement for a large number of runs, since the number of runs goes up as 2^m, where m is related to the number of factors being studied in the experiment (we will soon see what the relationship is). Thus, the concern with the subtopic mentioned above: ways to reduce the number of individual experiments needed to complete the experimental program successfully. The non-statistician approaches the problem in a manner that is rather simplistic, but actually works rather well under many conditions. That is, he simply allows one combination of conditions to represent the "base" or "reference" conditions, and then change one factor at a time to see what the effect is. Indeed, that is how most scientific experimentation is done. As we noted, it works satisfactorily under many conditions, yet it has its drawbacks. First, it implies that the experiments are carried out sequentially, which is bad if individual experiments take a long time. Secondly, it does not allow for an estimate of experimental error; the only way experimental error can be included in this design is to duplicate the experiments, and once you have done that, you might as well go to a more sophisticated design that will do it all more efficiently to begin with. One-at-a-time designs (as they are called) also fail to provide indications or estimates of interaction effects. For all these reasons, statisticians tend to shy away from this simplistic approach.

When we remember that the reduction of effort must be accomplished while keeping the design balanced, allowing for sufficient degrees of freedom for all the factors of interest to obtain good estimates of their effects, and making sure that each effect is measured with the best possible accuracy as well as including all the considerations that the one-at-a-time designs leave out, it is quite a feat to reach all these goals. However, statisticians have risen to the challenge and have come up with a good number of ingenious ways to manage this feat.

Again, we do not have space here to discuss all the different ways that have been devised (such as Latin Squares, which allows three factors to each be present at m levels, but only requires $m \times m$ individual experimental runs), but we will discuss one particularly ingenious design after noting the following generalities.

The biggest advantage of using all possible combinations of factors as an experimental design is that it allows all possible interactions to be noted and estimated. Thus, reducing the number of experiments requires that at least some of the interactions become confounded, either with each other or with the main effects. Thus, the details of any particular design become dependent upon which interactions are considered to be of interest and which may be known to be non-existent and therefore allowed to become lost in the interests of reducing an unmanageably large experiment to one that can be performed in a reasonable amount of time and with reasonable expense. By selectively eliminating certain combinations of factors, the experimental design can be reduced from the all-possible-combinations size by a considerable amount. These experimental designs are called "fractional factorial" designs, because the number of individual runs needed is a small fraction of the number needed for the full 2^m factorial experiment. As we just mentioned, exactly which runs to eliminate will depend upon which interactions are of interest, so that we cannot even begin a general discussion of how to set up such an experimental design. However, there is one particular design that is so nifty, and of such high efficiency and potential utility, that we cannot bear to leave it out.

This design is also nominally a fractional factorial design, but it has been reduced to such a small number of runs that its origin as a factorial experiment is all but lost. The basic factorial design has been pared to such an extent that *all* information regarding interactions has been lost. In exchange for this, we have gained the ability to include $n - 1$ factors in an experimental design that requires only n individual experiments. This is the same number of experiments that is required for the change-one-factor-at-a-time designs discussed above, which also lost all information about interactions, so where is the gain? The advantage to this design over the one-at-a-time experimental designs is that this design is completely balanced. This is another way of saying that the factors enter into the design independently. Investigation of the design matrix reveals that there are equal numbers of high and low levels of one factor for any level of another factor; and this relationship is true for any pair of factors in the design. A minor disadvantage of this design is that 2^m experiments are required even if not all $n - 1$ factors are being investigated. This reduces the efficiency of the design, although it still remains very efficient and is always balanced.

Table 29-1 shows, for two cases, how this design is set up. For the design in Table 29-1B, for example, consider factors 1 and 5. For two of the four experiments where factor 1 is at its high level, factor 5 is also at its high level; for the other two experiments where factor 1 is at its high level, factor 5 is at its low level. The same condition holds true for the four experiments where factor 1 is at its low level. Check it out: the same conditions hold for *every* pair of factors—the design is completely balanced.

The balanced nature of the design allows for several auxiliary advantages. For example, consider the disposition of the degrees of freedom. With eight experiments, there are only eight degrees of freedom for the response (the physical quantity being measured). One corresponds to the mean of the data, leaving seven for the seven effects being studied. This appears to leave no degrees of freedom for estimating either interaction or error. However, it is extremely

TABLE 29-1A
Three factors measured in four runs.

	Factor		
Run #	1	2	3
1	L	L	L
2	L	H	H
3	H	L	H
4	H	H	L

A very efficient design for experiments; $n - 1$ factors can be measured in n runs, as long as n is a power of 2. If the number of factors is not one less than a power of 2, then the efficiency decreases, but the design still remains highly efficient.

TABLE 29-1B

Seven factors measured in eight runs. If fewer than seven factors are of interest, then the number of columns in the design may be reduced as shown (in this case to five); however, the number of runs needed remains at eight.

	Factor						
Run #	1	2	3	4	5	6	7
1	H	H	H	L	L	L	H
2	H	H	L	L	H	H	L
3	H	L	H	H	L	H	L
4	H	L	L	H	H	L	H
5	L	H	H	H	H	L	L
6	L	H	L	H	L	H	H
7	L	L	H	L	H	H	H
8	L	L	L	L	L	L	L

unlikely that all seven factors will have significant effects. The differences between the levels of those factors that do not have significant effects on the response can be used to estimate the error; one degree of freedom becomes available for this purpose for every non-significant effect in the experiment—try *that* with a one-at-a-time design!

In Table 29-1 we show the design matrix for two sizes of experiment; but until you understand how the design matrix is generated, you cannot extend the design beyond what is shown here. The total number of experiments required is, as stated before, 2^m. m is chosen so that 2^m is at least one greater than the number of factors to be included in the experiment. Then the factors are numbered, and the first step is to generate all possible combinations of levels for m of the factors. For example, Table 29–1 shows the design matrix for $m = 2$ and $m = 3$; the first three columns of Table 29-1B have all possible combinations of the first three factors. For those of you who can count in binary, setting up these m factors should be trivial.

The next step is to generate the levels of other factors as follows: take a pair of the factors already set up, and generate the levels of the new factor by exclusive-or'ing the levels of those two factors. In other words, if the two existing factors have the same level in any given experiment, then the new factor gets a low; if the levels are different, then the new factor gets a high. This can be done for all pairs of factors set up in the first step. Thus when $m = 2$, one factor can be created this way; when $m = 3$, three factors can be created; and when $m = 4$, six factors.

The third step is to perform exclusive-or'ing between factors from the first step and factors from the second. In Table 29-1B, for example, the seventh factor was created by exclusive-or'ing the levels of factors 3 and 4. A key point to doing this is that a factor from step 1 must not have already been included in both the factors being used. Since factor 4 was made from the exclusive-or of factors 1 and 2, only factor 3 can be used with it. If we were to try factors 2 and 4, for example, the result would be the duplication of the levels of factor 1. (Exercise for the reader: create the design matrix for an experiment where m is 4, so that $n = 16$. Verify the balance of the design.)

Note that once m is chosen, 2^m must be at least one greater than the number of factors. If the number of factors is less than this amount, then the design can still be used, simply by ignoring some of the design columns. For example, if only five factors were needed in a particular experiment, then only the first five columns of the design matrix would be used, and the rest ignored.

The rationale for statistical designs of experiments is conveniently explained in terms of a prototype experiment. The prototype experiment is often called a "weighing" experiment, because it is conveniently put into terms of finding the weights of two masses. When this is done using a double-pan balance, there are two ways to determine the weights; one way corresponds to the standard experimental approach, one way to the statistical approach.

The standard approach is the one-at-a-time method: one mass is placed on one pan of the balance, and standard weights placed on the other pan until balance is reached; then the procedure is repeated with the other mass.

The statistical approach is to place both masses on one balance pan, and determine their combined weight, then place the two masses on the different balance pans and measure the difference in their weights. The weights of the individual masses are determined by calculations based on the two measured quantities.

This comparison of the standard versus the statistical approaches in the weighing experiment is useful because it illustrates most of the advantages and disadvantages of the statistical approach, in an experiment that is simple enough for everything to be clear. Each method requires a total of two weighings, and results in a value of the weight for each mass. In the statistical approach, however, the weighings are used more efficiently, in the sense that for the same amount of work, the weights of the two samples are known with greater accuracy. In this simple experiment, it is not possible to trade off accuracy for less work, but in more complex experiments, it is often possible to save a good deal of work, at the expense of increasing the error with which the final results are known.

The disadvantage of the statistical approach is that no information is available until all the data is collected and the computations performed. In general, there are no "intermediate results" available; all the answers pop out in the final stage of computations, just as the weights of the two masses only become available at the very end.

The last item we wish to discuss is the "exercise for the reader" that we brought up at the end of Chapter 27. Go back and review the full discussion leading up to the exercise; here we will present a shortened version of the exercise as a reminder before discussing the answer. The exercise asked the question: if an experimental design contained one factor with noise nested within the factor, but the data were treated as though it was a two-factor experiment, what would be the meaning of the results be if the F test for the noise turned out to be statistically significant?

The answer to this again goes back to the heart of statistics. For an F test to be statistically significant means that the result is too unlikely to have happened by chance alone. We must therefore assume that some systematic phenomenon occurred unbeknownst to us, and this previously unknown systematic effect repeated in the three experiments (i.e., the three levels of the factor). We cannot always tell *a priori* what that phenomenon might have been, although we can make some guesses. One likely possibility is drift; this would correlate with the order of collecting the readings and would likely be the same in all cases. Another possibility is an unconscious tendency of the experimenter (or technician who is running the experiments) to change his procedure in minor ways during a series of readings. Changes such as these need not be monotonic, as long as they repeat from one series of readings to the next. The point is that obtaining a statistically significant result is an indication that there is *something* at work that needs to be looked at. This is why experienced statisticians always do a hypothesis test, even in cases where logic seems to obviate the need for these tests. That was, in fact another "gotcha"—we put forth strong arguments to convince you that a hypothesis test need not be done in that case, when in fact it was not only appropriate, but mandatory. That is one of the key points: to avoid blundering into the pitfalls of statistics, hypothesis testing should be carried out routinely and should always be considered mandatory.

Here is where one of the more subtle traps for the unwary lies in wait. The novice statistician will read all our discussion so far and with all good intentions vow to always perform hypothesis tests, on a routine basis. Then, over a period of time, he will have performed several dozen (or even hundreds) of experiments, performed the hypothesis tests, and the first hundred or so came up as non-significant. Since, he then concludes, nothing ever shows up, there is little point in continuing to perform these tests so religiously, since it appears to be a waste of time. Then he stops. Guess what—this is when Murphy raises his ugly little head and shortly thereafter, during one of the experiments where a hypothesis test would have shown up an important result is when the test was not done. Complacency, or being lulled into a sense of security over the condition of the data, will trip you up every time.

We cannot devote as much space to a full discussion of all the possibilities of the needs that arise in an experiment that must be accounted for in the design, but we will briefly skim over some of them; interested readers can find further information in statistical texts. In addition to the conditions we discussed before, the nature of the sample determines the nature of the design. Often, when the conditions cannot be controlled, sampling designs must be used (as opposed to "experimental designs"), although there are similarities between the two concepts. Samples may be taken completely at random. Samples that are not completely random may be "stratified" by noting pertinent conditions. Stratification in sampling is similar to blocking in experimentation.

A key point to note here is that experimental designs are seldom "pure", especially when the experiment is complicated. Often both controllable and uncontrollable factors may be of interest, then some factors are crossed and some are nested in the same experiment. Sometimes some factors must be introduced by sampling, while others are controlled and made part of the experimental design. In all cases, a proper analysis of the data must conform to the way the experiment was set up and conducted. The initial partitioning of the sums of squares gives the correct breakdown of the data and shows which factors are determinable from the experiment and which not. The proper approach is to turn the situation around; the factors of importance should be specified in advance, then the correct experimental design can be constructed to insure that those factors will be included in the analysis (see Reference (29-2)).

References

29-1. Box, G.E.P., Hunter, W.G., and Hunter, J.S., *Statistics for Experimenters*, Wiley, New York, 1978.

29-2. Daniel, C., *Applications of Statistics to Industrial Experimentation*, Wiley, New York, 1976.

30

PITFALLS OF STATISTICS

Every now and then throughout this book we have indicated particular instances where the inexperienced, unwary, or careless data analyst can unknowingly stumble into problem areas and arrive at incorrect conclusions. We have also occasionally mentioned the cases where statistics have been deliberately misused in order to persuade through misrepresentation. Simultaneously we have kept giving warnings to "be careful" when analyzing data, particularly when doing what is loosely called "statistical" analysis of the data, because of the high probability of falling into one of the traps. On the other hand, a warning to "be careful" is of little use if you do not know what to be careful of, how to recognize it, and what to do about it if found. In one sense, almost the entire mass of the published literature in statistics is just that: a compilation of the problems encountered, and the solutions to those problems, when analyzing data.

That does us little good here; we could hardly reprint the entire statistical literature even if we had a mind to and copyright problems did not interfere; we need to find a middle ground. Most of the papers in the statistical literature, just as most of this book, are written in the positive sense: the emphasis is on telling you what to do, and the information as to what to avoid is scattered within the mass of the other information.

We have, however, come across some books whose main subject is just that: the exposition of the problem areas of statistics (30-1 – 30-3). The title of the book by Hooke in particular is indicative of the attitude, and shows that we are not the only ones who recognize the existence of the problems involved; furthermore, all of these authors are not only trained statisticians, but each also has the expanse of an entire book with which to present the topic—they are all highly recommended. The point, of course, is that there are not only those who deliberately misuse and abuse the power of statistical calculations, but also those who, while well-meaning, are too inexperienced to recognize or guard themselves against the statistical traps that abound, and so lead others into the morass. Both Hooke and Moore have created names for such people: Hooke calls them "data pushers" and Moore calls them "number slingers". These data pushers generally tend to be convincing because the half-truths that they disseminate are usually infinitely more dramatic than the real truth, so before you can say "yes, but…", everyone has been convinced and the false information has probably already appeared in a newspaper.

Yet there is a difference between the approaches of the books. Hooke, as you may be able to tell by the title, concentrates solely on the negatives (i.e., the pitfalls), and how to detect them; Moore presents the material half-and-half: half positive as to what to do, and half negative as to what to watch out for. However, even when discussing the positives, some of the discussion deals with how to use statistical tools to examine data, to test it for internal and external consistency, for relevance, completeness, etc. Jaffe and Spirer compare improper to correct methodology.

Fortunately, the existence and nature of misuse of statistics is finally starting to become more widely disseminated; an article dealing with this topic has even appeared in *Scientific American* (30-4)

After examining these books, we have combined their comments with our own experience and compiled a list of potential pitfalls that need to be guarded against; this is in addition to those pitfalls that we have indicated now and again in earlier chapters. Jaffe and Spirer also compile a list and categorize misuses into five general categories:

(1) lack of knowledge of the subject;
(2) quality of the basic data;
(3) preparation of the study and report;
(4) statistical methodology;
(5) deliberate suppression of data.

Our list is more detailed and extensive, and even so we feel is incomplete. On the other hand (count the hands: is this a misuse of statistics?), several of the possible pitfalls may not apply to the particular uses that we, as spectroscopists, intend to put statistics. Statisticians as a group have to deal with all kinds of situations; some of the more exotic pitfalls are encountered when the nature of the experiment (or survey) involves experimental units that are either self-aware or aware of the existence and/or purpose of the experiment. Animal subjects can be cantankerous, to say nothing of the problems encountered when that ultimate in contrariness: other human beings, are the experimental units. Fortunately for us, those problems are encountered mostly in experiments in the medical and social sciences—however, various spectroscopic techniques are more and more coming to be used in conjunction with such applications: beware. Besides, every experienced scientist knows that even supposedly inert laboratory equipment often acts as though it is maliciously aware of the experimenter—or are we falling into the trap of anthropomorphism? (Exercise for the reader: is that a statistical or a non-statistical trap?)

One could argue that the list is really too long; that some of the individual traps we include are really different manifestations of the same underlying phenomenon, such as misuse or misinterpretation of some particular statistical tool. Well, that may be, in some cases, but we will not argue about it: this is our book and we think the list is about right.

Well, here is our list of over 30 statistical traps to watch out for, along with our comments about them; because of the extent of the list, the commentary may be a bit sketchy in places.

(1) The first item on our list is the fact that numbers seem to have a life and a mind of their own. There is a situation that can arise and, in an incident that has become famous among statisticians, actually did arise at a well-known university accused of discriminatory practices. Table 30-1 presents data that are hypothetical, but of the type that led to the problem (both the data and the department names are synthetic and are chosen simply to illustrate the nature of the problem). It is clear that both the history department and the mystery department are admitting the same percentage of "discriminated" students as all

TABLE 30-1

Data illustrating how numbers themselves can lead to erroneous conclusions, depending on how they are grouped. Since the actual situation occurred at a university, we label the data with labels appropriate to that venue (although not the actual ones in the real case).

	No. of applicants	No. admitted	% admitted
History department			
Discriminated group	10	5	50
Others	20	10	50
Mystery department			
Discriminated group	20	6	30
Others	10	5	30
School totals			
Discriminated group	30	11	36.6
Others	50	13	43.3

other students (so that the number admitted from each group is in proportion to the number that apply). Yet since the history department is larger, it can accept a greater number of students, and since it is apparently more popular among the "other" students than the mystery department is, the total number of students admitted in the entire school, as well as the percentage, is lower for the "discriminated" students than for the others. There is no cure for this situation; it is a real property of the numbers involved, and has come to be known as "Simpson's paradox". This sort of stuff is tantamount to an announcement of open season for the number slingers, who can present the data in whichever way best suits their purposes. All that can be done is to be aware of the existence of this behavior of data, so that when it arises it does not lead to unfounded claims (mathematical or otherwise).

(2) Another real effect, caused by the properties of some situations, is the non-transitive relations can exist among three or more groups. Transitivity is the property that says that if A and B are related in a certain way, and B and C are related in that way, then A and C are also related in the same way. For example, the magnitude of numbers is a transitive property: if A is greater than B and B is greater than C, then we can legitimately conclude that A is greater than C, because of the transitivity property. However, not everything that can be measured is transitive. The classic situation that arises has to do with election data, and people's responses to different candidates. The situation is complicated to describe, but if a minority of the voters prefers Candidate A strongly, and a majority prefers Candidate B weakly over Candidate C, while a different minority prefers Candidate C strongly, the stage is set for a non-transitive set of voting preferences. Then, if A can win over B and B can win over C, there is no assurance that A can win over C, if only those two candidates are running. The danger here is that of extrapolation, a subject we will have more to say about further on.

(3) Large populations are more likely to have extreme individuals than small populations. For example, if 100 units are in the sample, then we expect five to be beyond the 95% confidence limits. In the case of spectroscopic calibration procedures (we are getting a bit ahead of ourselves here, since we have not yet discussed calibration), the residuals from the calibration are often examined, with a view toward eliminating "outliers". All to often, a fixed cutoff point is set, and any observation that is beyond the cutoff is eliminated, despite the fact that, for large enough assemblages of data, it is virtually inevitable that *some* data will be beyond those limits by chance alone and are fully valid parts of the data.

Similarly, when many measurements are taken, such as in process control situations where data are taken continuously, sooner or later a value at one extreme or the other will appear. Now, process control is an interesting situation, because it is usually desired to maintain the average at a given value. A common situation is that a change in one direction will violate either a regulatory or industry standard (or simply make the product defective for its intended use), and a change in the other direction costs money. Hence, it is every bit as important not to change the process whenever an extreme chance value comes along as it is to correct for real variations in the product. Furthermore, as we have shown in our discussion of synthesis of variance in Chapter 4, every additional source of variation adds to the total; therefore, adding variance by changing or adjusting the process unnecessarily will in fact increase the total variance of the process conditions.

(4) Failing to perform hypothesis tests. We have discussed this in connection with ANOVA calculations; if a component of variance is calculated, it must be shown by a hypothesis test that it is due to a real phenomenon; otherwise the random variations due to the noise will be interpreted as being meaningful (an event we have seen all too often, and clearly an incorrect result). This is true, of course, in general, not just for ANOVA calculations. All the discussion we have presented dealing with hypothesis testing has been toward the goal of making our readers realize that simply assuming that the results are real (without testing them) is one of the pitfalls, and that hypothesis testing is the way to test the assumptions.

(5) A related pitfall is to perform hypothesis tests, but become discouraged when many of the effects turn out to be non-significant, and stop testing. We discussed this recently and will not belabor the point now.

(6) Failing to test for randomness: assuming that the part of the data that are supposed to be random actually are, without testing them. A similar pitfall is assuming that the data follow some particular distribution without testing them. At the least, this will lead to incorrect confidence limits for the actual (unknown) distribution and thus incorrect conclusions even when a hypothesis test is carried out. When randomness is assumed for a particular statistical calculation (and most statistical calculations are based on *some* quantity being randomly distributed) and randomness does not exist, then of course that calculation should be thrown out. Unfortunately, this is rarely done because the randomness of the quantity in question (e.g., the residuals of a calibration or an ANOVA) is rarely tested, so that when randomness is not present, that fact remains unknown.

(7) Assuming that correlation indicates causation. There are at least two reasons that cause variables to be correlated besides true causation of one by the other. These reasons are: (a) confounding of variables (of which we will have more to say further on) and (b) both measured variables are responding to the true (perhaps unknown) cause of both.

In any case, correlation, as measured by the correlation coefficient, is a statistic. As such it is amenable to being subjected to hypothesis testing. Tables of the critical values of the statistic r (sample correlation coefficient) for values of the population correlation coefficient $(\rho) = 0$ are readily available (30-5); tables for other values of ρ can also be found, although not as readily.

(8) "Fallacy of the unmentioned base"—this one is apparently popular enough among number slingers to merit a formal name of its own. Basically it amounts to choosing the most dramatic way to present data (as opposed to the clearest way), so that while the underlying data and results are correct, they are presented in such a way as to mislead. Thus, if a small percentage of a large number of possible disasters actually happens, the absolute number of disasters may be given instead of the low rate. Alternatively, if a high percentage of cases may be disastrous (e.g., a high mortality rate for a rare disease), then the percentage may be emphasized rather than the unlikelihood of getting the disease—it is more dramatic that way. The cure is to make sure that you know both numbers, or, even better, a full description of what the numbers refer to—the "base" that is not being given voluntarily.

Examples abound. A recent article in my local newspaper stated "...50% of marriages end in divorce...". Even to my wife, who is completely non-technical, this did not sound right: "...hardly anybody we know has ever been divorced...". I then pointed out to her the other fairly well-known fact that most divorces occur to the same people, therefore those few people who marry and divorce multiple times add disproportionately to the numbers. As a simplified example, consider 1000 couples, 250 of whom marry and divorce three times, and the rest remain married to their first spice (plural of spouse?—mouse, mice, spouse, spice?). In any case, these 1000 couples account for 1500 marriages, and indeed, fully half of those marriages ended in divorce, even though only a fourth of the people involved were ever divorced—but which makes the more dramatic newspaper story?

(9) "Other things equal"—this covers a multitude of sins. Are all other things really equal or not? Unless the data were taken using a carefully balanced statistical experimental design, or a properly designed and executed sampling survey (often necessary in social science applications), other things are usually not equal. The most common experimental design that makes "other things equal" is the use of "controls" in the study; the idea of a control is that "other things" than the factor under study are kept constant. Yet, despite all efforts, that may not be so (and worse, the failure of other things to be equal may not be apparent). Indeed, it is sometimes found that even when statistical designs are used, the existence of an important factor was not suspected and therefore not taken into account in the design—in which case it is back to the drawing board. "Other things not equal" is often (although not always) equivalent to confounding of factors in the design.

(10) Obviously, the next item is the confounding of factors—the distinction from "other things equal" is a little nebulous. Indeed, there is a good deal of overlap between these pitfalls. Basically, this is a little more general; confounding of other factors with the desired one can occur whether or not extraneous factors are controlled; when extraneous factors are not equal but their effects are known,

their contribution to the result can be taken into account via calculation rather than by cancellation in the experiment. In fact, more applications of science use this approach than the "pure" statistical approach, although the basic idea of using controls is applied extensively in this type of approach also.

(11) The belief that since probabilities converge in the long run, there is a "force" that tends to even things out. This belief apparently is especially pernicious among gamblers, who fail to realize that every new bet is the beginning of a new series that has nothing to do with the previous bets.

In fact there is a rather peculiar relationship between the convergence of probability and the actual number of cases at the convergence value. Consider a simple situation, such as the binomial distribution. In long series of tests (say, coin tossing), the probability approaches 50% heads (or tails), and as the series of tosses gets longer, the average number of heads will get closer and closer to 50%. Yet, the probability of tossing exactly 50% heads decreases continually. There is a single underlying reason for both these observations, and that is the proliferation of possible outcomes for the sequence of tosses. Since coin tossing is a discrete event, we could (in principle, at least) count all the cases. What we would find is that more outcomes arise that are close to 50% than are farther away; at the same time, the number of outcomes that represent exactly 50% increases much more slowly than the total number of outcomes, thus reducing the probability of attaining exactly 50%. As an illustration, suppose we toss a coin three times; the proportion of cases (which estimates the probability) can never be closer than 17% away from the 50% mark. But if we toss the coin five times, we can get within 10%. In neither of these cases can we obtain a result of exactly 50% no matter what the sequence of tosses is.

(12) In many, if not most, situations both systematic and random effects operate. All too often, the results of the random effects are assigned to the systematic effects, giving the appearance of real differences where none exist. This situation is exactly what hypothesis testing is intended to tell us the truth about. Unfortunately there are limitations even to hypothesis testing. For example, in sports (and other competitive activities) the real difference in ability between the top few players is probably negligible (we hedge because we cannot really prove it any more than anyone else can), but clearly there are always random factors that affect who wins a given tournament (state of health, state of mind, the weather will affect each player differently, etc.). In these terms the systematic and random factors are called "skill" and "luck", respectively. To say that player A is more skillful on Monday because he won on Monday, but player B improved enough to be better on Tuesday because he won on Tuesday, is obviously nonsense; the difference is due to the random factors or, if you will, their "luck" changed on the different days.

Even in physical situations, it is not always clear how much of the results are due to random perturbations; the system may not have been studied sufficiently, and the distribution may be unknown. One example is a situation that we have studied recently, and deals with the question of wavelength selection in spectroscopic calibration. For a long time, computer-selected wavelengths were assumed to represent the result of real underlying spectroscopic effects, until a Monte Carlo study revealed the extent to which the random noise of the spectrometer affected the selection process (30-6). Knowledge in this area is still not complete, however; the exact analytic form of the distribution function has not yet been determined (good exercise for the mathematically oriented reader, but it is definitely a research project and if anybody solves it please let us know).

Our list is long enough that we split it up and will continue it in the next chapter, right from where we left off.

References

30-1. Hooke, R., *How to Tell the Liars from the Statisticians*, Marcel Dekker, New York, 1983.

30-2. Moore, D.S., *Statistics: Concepts and Controversies*, W.H. Freeman & Co., New York, 1985.

30-3. Jaffe, A.J., and Spirer, H.F., *Misused Statistics—Straight Talk for Twisted Numbers*, Marcel Dekker, New York, 1987.

30-4. Dewdney, A.K., *Scientific American*, 274: 3, (1990), 118–123.

30-5. Owen, D.B., *Handbook of Statistical Tables*, Addison-Wesley, Reading, MA, 1962.

30-6. Mark, H., *Applied Spectroscopy*, 42: 8, (1988), 1427–1440.

31

Pitfalls of Statistics Continued

(13) There are two beliefs about the behavior of random numbers, which, even though they are contrary to each other, are both held by various people (and, we suspect, are sometimes both held by the same person). The first one is called the regression fallacy; this is the belief that there is a tendency for populations to tend to regress toward the mean as time goes on. This arises as follows: assume a population with known values for the mean and variance. To put the fallacy into the terms in which it was first noted historically, consider the heights of men in a given region, in which case the variable is Normally distributed. Now, from the shape of the distribution alone, it is clear that for a tall man, the probability of his son being shorter than himself is over 50%, while for a short man, there is a greater than 50% probability of his son being taller than himself. At the time this was noticed empirically, it led to the belief that the population as a whole tended to regress toward the mean, and that very tall and very short men would eventually disappear. The fallacy of course, is that even though few tall or short men are born, there are some, just enough to replace those that die, and so maintain a stable distribution of heights in the population.

The opposite belief is the one that ignores the fact that extreme values of the population exist and are more and more likely to occur as the population size increases. For example, with a sample of 100 items, there should be 5 beyond the 95% confidence limit; the probability of having less than 2 beyond the 95% confidence limit is very small. (Exercises for the reader: what distribution will the number of data points beyond the 95% confidence limit follow? What is the probability of having less than 2 beyond that limit?) If the sample size is 1000, there should be 50 beyond the 95% confidence limit, and 10 beyond the 99% confidence limit; it is inconceivable for there to be none. Yet there is a tendency to believe that, despite the sample size, any measurement beyond the 99% limit, or even the 95% limit does not belong, that it must be an "outlier" or some such, and should be rejected from the set. Now, the answer to the question of what constitutes an "outlier" is not at all clear, even among statisticians, although it could be loosely defined as an observation that does not belong to the target population. However, defining an outlier is one thing, but determining if a given datum is an outlier is something else entirely; declaring a reading to be an outlier simply because it lies beyond a given confidence level is clearly too simplistic and too likely to be wrong, especially when there are many observations.

(14) Assuming independence when it does not exist. Everybody has experienced "bad days", when everything seems to go wrong. If all the disasters that happen on a "bad day" were truly independent, then the probability of them all occurring at once would indeed be so minuscule that "bad days" would be virtually non-existent. A close scrutiny of the events that happen on a bad day would reveal that one disaster contributes at least partially to the next, so that they are not independent, and consequently not all that unlikely. Formal testing of such phenomena is virtually impossible, because the probabilities of the individual events are unknown to start with, and "bad days" occur rarely enough that statistical tests are useless. However, examining the physical causes can reveal the connections. This is the other side of the coin of believing that correlation implies causation (see no. 7): here the belief in lack of causation gives rise to the belief in lack of correlation, and leads to a belief that the probabilities of the combination of events is much lower than they truly are.

In scientific work, the situations are more amenable to objective testing, but too often proper hypothesis tests are not applied; more attention to this aspect of experimentation is warranted.

(15) Misunderstanding the relationship between type I and type II errors (as the statisticians call them). We have declared time and time again that "only" 5% of the data will lie beyond 95% confidence limits. However, if you are unlucky enough to get one of those 5% during a hypothesis test (and it can happen, indeed it *will* happen, 5% of the time), then you will incorrectly conclude that a significant effect is present. This is a type 1 error. A type II error is the opposite: concluding that no effect is present when there really is one. This can happen when the effect is so small that its distribution overlaps the null distribution, and we are again unlucky enough to take our reading on one of those few observations that lead us to the wrong answer. The problem with type II errors is that they depend upon a distribution whose characteristics are unknown, so that we usually cannot define the probability levels—after all, if we knew the characteristics of the alternate distribution (remember, that represents the population that our

measured data comes from), we would not need to do a hypothesis test in the first place since we would already know the answer.

Thus, for example, we can extend the confidence limit to reduce the probability of a type I error. If we do that, then unless the alternate distribution is so far away from the null distribution that confusion is impossible, inevitably we increase the probability of a type II error. [Exercise for the reader: demonstrate this by drawing the null distribution (i.e., the one corresponding to the null hypothesis) and several alternate distributions. What happens when the distributions overlap, and when they do not overlap?]

(16) Confusing zero with nothing. In Chapter 13 we discussed this situation in connection with what happens when you run out of sums of squares and degrees of freedom at the same time. What you are left with is no information regarding the variability of the data, a far cry from having data with no variability. A similar situation exists, for example, in the batting average of a baseball player who has never been at bat. However, equating nothing with zero (by simply assuming no variability in the data) is apparently often done inadvertently in more serious cases when "something must be used" in the absence of information.

(17) Assuming linearity (without checking). Some things improve with magnitude (e.g., economies of scale in manufacturing) while some things get worse (e.g., the post office—and never mind comments about the post office getting worse no matter what). However, the belief that a situation will scale up without change is generally unfounded.

(18) Polls and opinion surveys are fraught with the potential for problems. Fortunately, they have all been encountered in the past (well, all the ones that are known, and let us hope there are no major surprises left). Trained statisticians are aware of them, and can therefore (hopefully at least) avoid them. Bad experiences with early survey methodologies almost led to the demise of the practice, and reportedly did lead to the demise of at least one magazine practicing it in the political arena (*Literary Digest*, whose poll predicted Alf Landon winning over F.D.R. in 1936).

One major problem is lack of money. This results in insufficient resources to overcome the other problems. However, one school of thought holds that even a poor survey is better than none at all.

Biased questions—the wording of the survey questions—have been known to affect the results. This is especially true for controversial issues, but can happen in any survey.

Poor or incorrect sampling—the sample will not represent the target population. Two of the blatant problems are convenience sampling (e.g., passers-by on the street) or self-selection sampling (in a mail survey, the sample represents only those who feel strongly enough about the topic to bother filling out and returning the survey form).

Sample too small—this leads to excessive sampling error. However, this can be mitigated by computing the confidence limits for the size sample used. One case to watch for especially is if exactly 67% of the samples showed an effect (or no effect)—there is a good chance that only three samples were used.

Changes in the population—this is a problem particularly noteworthy in election polls, where the voters may change their minds between the poll and the election.

(19) Confusing statistical significance with practical significance—this is the other side of the coin of the need to perform hypothesis tests. A statistically significant result merely tells us that a difference exists; it tells us nothing about the magnitude or practical importance of the difference. This is a favorite ploy of the "data pushers", who can then claim to have used statistics to prove their case.

(20) Use of the double negative—again, the opposite of no. 19. Failure to prove a difference does not mean that there is no difference—our experiment may not have been sensitive enough (i.e., may have been too small—recall that the width of the confidence limits for $\bar{X}$ decreases with the square root of n; a larger n may reveal the difference).

(21) Confusing cause and effect (or ignoring a third, correlated factor)—does being an athlete help one live longer? Or do good health and regular exercise do both? (Exercise for the reader: if you find out for sure, let us all know!)

(22) Extrapolating or generalizing the results beyond the confines of the experiment. Most scientists know the danger of extrapolating a calibration curve beyond the range of samples used in the calibration, but many are blissfully unaware of the similar danger of extrapolating any result beyond the population for which its validity is verified. Small extrapolations are often used (otherwise the term "learning from experience" would be non-existent), but as in the calibration case, the greater the extrapolation, the greater the danger of error.

(23) Failure to use adequate controls—a controlled experiment will be on the order of twice the size of the same experiment without controls. This is a temptation to avoid the controls (see also no. 9: "other things being equal").

(24) Unconscious bias on the part of the experimenter—it is a well-known fact (among statisticians, certainly, and most other scientists also) that the experimenter can unconsciously affect the results of the experiment (and no quantum mechanical considerations need be brought into play). Particularly when the subjects are human, as in medical experiments, knowledge on the part of the patient as to which treatment is being used can affect the response. This is sometimes called the "placebo effect", and is counteracted by the use of "blind" experiments, where the patient does not know which treatment is being used. Even in blind experiments, however, it has been found that the attitude of the scientist (or physician) can affect the subject's response to the treatment. This is counteracted by use of "double blind" experiments, where neither the physician nor the patient knows which patients receive which treatment. This type of

double-blind experiment should not be confused with the type espoused by the late Tomas Hirschfeld who, among his other accomplishments, claimed to have done a double-blind experiment all by himself—because when he had to analyze explosives at Lawrence Livermore Laboratory he closed *both* eyes!

(25) Sloppiness—even well-designed experiments can be ruined by poor execution, or by failing to avoid the other pitfalls that statistical calculations are subject to (see Reference (31-1, p. 81) for an example of an important experiment brought to almost complete disaster by carelessness during its execution).

(26) Ethics and humanitarian actions—these types of behavior, while both highly laudatory, can play havoc with the information that might have been obtained from an experiment. Sometimes ethics (especially medical ethics) may prevent certain experiments from being done at all, while other experiments may be ruined (at least from the point of view of obtaining unambiguous data) by a solicitous experimenter giving the treatment that is expected to be "better" to those who need it most (i.e., are sickest). This type of humanitarian behavior inextricably confounds the treatment with the patient, and it becomes impossible to tell if the treatment really worked, or if the sicker patients recovered faster naturally, or if the sickest patients responded to the *knowledge* that they were receiving the supposedly "better" treatment, rather than to the treatment itself. Double-blind experiments are needed here, also.

(27) Graphs can mislead, if the scales are inappropriate. Small effects can be made to appear major by stretching the axis, and vice versa. Careful inspection of the scale should expose such manipulations.

(28) A number given as a measure of variability is meaningless unless an exact statement is made as to what the number represents. For example, if the error of a certain analysis is stated as 0.01%, can you tell which of the following terms that error represents:

(A) Maximum error
(B) Peak-to-peak error
(C) One standard deviation
(D) Two standard deviations (etc...)
(E) Error at some stated confidence level.

We cannot.

(29) Finite population distorts probabilities—this is a subtle problem that may arise during simulation studies. If sampling without replacement is performed, then the probabilities change with every sample, and the exact probability depends upon all the previous samples. For example, if a container of 50 red and 50 black balls is sampled, the probability of either one is 50/100. If a red one is chosen, then the probability of red on the second draw is 49/99, and of black is 50/99—the probabilities are no longer the same.

(30) For some reason, the correlation coefficient has assumed undue importance: any non-zero value is assumed to indicate a real relationship. As we mentioned before, however, it is only a statistic like any other statistic, and must be compared against confidence limits before valid conclusions can be drawn.

(31) A highly significant result does not mean strong association, but rather strong evidence of some association. This is akin to the distinction between statistical and practical significance (see no. 19). Here, however, we also note that the distinction between "significant" and "non-significant" is not a sharp cutoff line, beyond which we are certain that all effects are real, and below which all is due to chance, but rather recognizing that we have more and more reason to believe in the reality of the effect, the higher the confidence level at which we can measure it.

(32) Do not test a hypothesis on the same data that suggested that hypothesis. At least two possibilities can arise: first, one set of data can show significance due to chance alone, so if the hypothesis was formed *a posteriori* based on that data, then it will certainly test out as being significant. Second, retrospective examinations of historical data may show previously unsuspected relationships. Again, these relationships would not have been noticed if they were not strong enough to be obvious, in which case they will again test as being significant. In both cases the proper action is to use the existing information as a guide to setting up an independent and properly designed experiment to test the hypothesis.

(33) Performing more than one hypothesis test on a set of data. If each test is performed at the 5% level, then the probability of at least one test being significant is $1 - 0.95^n$ (see Chapter 2 for more about how to compute probabilities). It takes only 14 tests for the probability to rise to over 50% that at least one will be significant by chance alone. If it is absolutely necessary to do more than one test, then do not consider a result significant unless it is significant at a level of $0.95^{1/n}$ (where n is the number of tests to be done). (P.S. Notice how we emphasize our own argument by using loaded words: "*only* 14 tests"—does 14 rate an "only"?)

(34) Believing the average covers all cases—this is closely related to pitfall no. 3. Instead of being surprised at the existence of data beyond some given confidence limit, such cases are simply ignored. Even maintaining the average may not be sufficient. The example here comes from queuing theory (described by Hooke (31-2, p.47)). If a clerk can process a customer in a minute, and one customer arrives every minute on the average, what will happen? It turns out that since the customers arrive at random intervals, even though the average arrival rate is one per minute, the clerk will lose some time in those cases where the interval between arrivals is longer than a minute. This lost time can never be made up, and so the line will tend to grow indefinitely (and the fact that some customers will also lose indefinite amounts of time waiting on the line does not compensate).

(35) Confusing the sources of error—to get ahead of ourselves again, we have seen situations dealing with calibration data where "improvements" in the calibration performance were spurious at best. It is possible to apply the formalism of partitioning the sums of squares to separate the regression error into three parts: error in the dependent variable, error in the independent variable, and lack of fit. At least in theory, the errors in the two variables can be ascertained by a suitable experimental design, but the lack of fit cannot be directly determined. "Improvements" in calibration performance have often been claimed in cases where it is not clear which error was being improved or whether a reduction in random error was achieved at the expense of lack of fit.

(36) Overdependence on the computer, resulting in the inability to perform simple tasks, and thus becoming limited to the set of tools available in a particular software package. Often, useful plots of the data are simple enough to be drawn by hand, and it is good practice to do so. Over the years, statisticians have created many pencil-and-paper tools for attacking the problems endemic in data. They did it of necessity; but the knowledge exists and can be used. It is not possible for any manufacturer of equipment, or even all of them together, to create computer programs for every possible task; the alert data analyst will seek out such tools in the literature and apply them himself, with or without a computer. Plotting and otherwise examining data by hand also provides an intimacy with the data that is otherwise impossible to achieve.

The careful reader will have noted that most of these pitfalls involve failure of one assumption or another to hold. Some of the most pernicious involve assumptions regarding randomness and/or distributions. The way to avoid falling into these traps is to test all assumptions, both explicit and implicit, before continuing on with the experiment or its analysis. A key word here is "all"—if an important assumption is missed, the result could be one form or another of havoc. In pitfall no. 4, we mentioned the need to perform hypothesis tests; here we emphasize that rather than just accepting them, hypothesis testing is the way to test the assumptions. It is all too tempting to test only the obvious or easy ones, but that does not suffice.

References

31-1. Moore, D.S., *Statistics: Concepts and Controversies*, W.H. Freeman & Co., New York, 1985.
31-2. Hooke, R., *How to Tell the Liars from the Statisticians*, Marcel Dekker, New York, 1983.

32

CALIBRATION IN SPECTROSCOPY

Now we come to what is probably the most confusing and difficult topic that chemists/spectroscopists have to deal with: the problem of calibrating spectroscopic instruments to do quantitative analysis. It really should not be that way; everybody has calibrated something at one time or another, even if only as a laboratory exercise in an undergraduate course. However, besides the real complications that arise when calibrating instruments on "real world" samples, there is an enormous amount of confusion sown by conflicting claims made by anyone and everyone who has their own "magic answer" to all the problems that arise, and furthermore has packaged it in a computer program for which "all you have to do is push the buttons".

To some extent, we ourselves have a problem in deciding how to deal with this frustrating and exasperating situation. Statisticians have, indeed, done wonderful things in terms of developing methods of dealing with the many real problems that can and do arise during the calibration process. Of course, for the most part they did not develop them for the purpose of calibrating instruments; they often did it just as a general method of solving some particular type of problem in any environment. Sometimes it was in response to a particular need, but even in this case, the method would be generalized before being published in the open literature. Thus, there exists an enormous body of knowledge just waiting to be applied to our spectroscopic problems. The problem that we (the more general "we", the one that includes you, our readers) as chemists have is to pick and choose from this vast knowledge bank those pieces that are likely to be of use to us. The particular problem that we (the more specific "we") have as authors is not only to decide which pieces to present, but also to figure out how to present it in our own inimical (inimitable?—well, something like that) style, presenting these complicated topics in a simple way (so everyone can understand it), while still retaining the full mathematical rigor (so that everybody will believe it), without using any mathematics (so nobody will be scared away), and, as is our wont, doing it all in words of one syllable or less. And you thought you had problems!

Well, we cannot do all of that, much as we would like to; it is simply too vast a topic. The statistician's approach to calibration is, in general, regression analysis in its various and varied forms. Regression is treated in virtually every book dealing with elementary statistics, and there are entire books devoted solely to that topic (see References (32-1–32-4)). Reference (32-1), in particular, is the type of book that is classic, and should be considered required reading for anyone with more than the most casual interest in regression analysis or calibration. Books for chemists, dealing solely with calibration considerations for chemical analysis, are rare. One such book was written by one of us (HM) dealing with this topic (32-5).

Other than that, we will try as well as we can to approximate the list of qualities we want our book to have, in the brief space available for them; we will be discussing the question of calibration considerations for the next several chapters.

We will begin by noting that the general purpose of calibration is to develop a mathematical model that relates the optical data from an instrument to the concentration of an analyte within a specimen consisting of a mixture of materials. Our first topic of interest is thus entitled:

Relating Absorbance to Concentration

Algorithms used to "interpret" optical data for absorbing samples may be explained as different approaches to relating optical data at specific wavelengths to analyte concentration via Beer's Law. To put this in mathematical terms that should be familiar to chemists, and certainly to spectroscopists:

$$A = MCd \tag{32-1}$$

where

A = absorbance (optical density, now an obsolete term, is sometimes found in older texts),
M = molar absorptivity (sometimes ε is used for M. Another obsolete term, used in older texts for absorptivity, is extinction coefficient),
C = molar concentration of absorber,
d = sample pathlength,

and by simple algebraic transformation:

$$C = A/Md \tag{32-2}$$

In simplified notation, we can assume unity pathlength, and then note that the Beer's Law relationship becomes:

$$A = \varepsilon C \pm E_A \tag{32-3}$$

where E_A stands for the error in A, and thus:

$$C = \varepsilon^{-1} A \pm E_C \tag{32-4}$$

where

C = concentration of the components giving rise to each spectrum,
ε = extinction coefficient for the specific component at a given wavelength,
A = absorbance of a sample for a specific wavelength or set of wavelengths, and
E = unexplained error associated in describing concentration as a function of absorbance (this is what happens in real life, and the reason regression must be used to create the calibration model, rather than simply solving simultaneous equations).

In more complicated cases, where interferences may exist, a multi-wavelength regression equation must be used; this is commonly expressed for calibration as:

$$Y = B_0 + B_i \times \log\left(\frac{1}{R_i}\right)[+E] \tag{32-5}$$

where

Y = percent concentration of absorber,
B_0 = intercept from regression,
B_i = regression coefficient,
i = index of the wavelength used and its corresponding reflectance (R_i),
N = total number of wavelengths used in regression, and
E = random error.

This is actually a form of Beer's Law with each B term containing both pathlength and molar absorptivity (extinction coefficient) terms. We place the error term, E, in brackets to indicate that it may or may not be included in the formulation of the model. Including this term results in the most general model, but it is also more difficult to solve and gives rise to the requirement of using a statistical approach to the situation, rather than a simpler direct physical formulation. The inclusion of the error term in the model permits solutions to be generated in cases where the simpler direct physical model will fail.

However, we will start with the simpler case; in its most simple physical formulation, the concentration is related to the optical data as:

$$\text{Conc} = \frac{\text{Change in Concentration}}{\text{Change in Absorbance}} \times \text{Absorbance} + \text{Some Error}$$

or, Conc $= B \times$ Absorbance + Some Error.

Thus B, the regression coefficient, is equal to the change in concentration divided by the change in absorbance [and, by the way, has units of (concentration units per absorbance units), e.g.: %/absorbance]. Therefore, if there is a large change in concentration for our calibration set with a relatively large change in absorbance, our regression coefficients tend to stay small, indicating a high sensitivity and good rejection of noise. In contrast, if we have a large concentration change relative to a small change in absorbance, our regression coefficients tend to be large and indicate low sensitivity and low signal-to-noise. This simple viewpoint applies whether a univariate (one wavelength) or multivariate (many wavelengths) regression is performed.

If calibrations are performed for the simplest case involving one analyte per wavelength and no instrumental error (i.e., no noise or instrument drift with time), Equation 32-3 could be used with a single term:

$$Y = B_i \times \log\left(\frac{1}{R_i}\right) \tag{32-6}$$

where the concentration is represented by the mathematical symbol Y and the absorbance is represented by the logarithmic term. We use the symbol R by habit, since that represents reflectance and our background happens to be intimately involved with reflectance measurements. However, there is a curious dualism that we can use to our advantage: the term "absorbance" is used in different branches of spectroscopy to mean the negative logarithm of reflectance, or the negative logarithm of transmittance. However, since the mathematical formulation is the same for the two cases, every mathematical expression is equally valid for both meanings of the term "absorbance", and therefore can be interpreted and used the same way regardless of which branch of spectroscopy is involved.

By knowing Y, and measuring $-\log R$ at each wavelength i, Equation 32-4 could be used to solve for multipliers (B) of the absorbance terms. However, in the real world, complications exist, such as instrument noise, drift, non-linearity between optical data and analyte concentration (deviations from Beer's Law), scattering, non-linear dispersion, error in reference laboratory results, physical property variations in samples, chemically unstable samples, band overlap, band broadening, and sampling errors. Equation 32-3 may be used to solve for B_0 and

B_i, if ideal conditions are approximated, namely:

(1) noise is assumed to be stochastic (random) and not unidirectional,
(2) precautions are taken to optimize signal-to-noise ratios,
(3) reference readings at each wavelength are collected to negate drift considerations, and
(4) excellent laboratory technique and sample selection protocol are followed.

This multilinear regression is a mathematical model relating the absorbance of several samples to their analyte concentration [assuming the analyte concentrations have been previously determined via an independent, accepted reference (usually a wet chemical) method]. The regression provides a best-fit linear model for absorbance versus analyte concentration, and the mathematical model used minimizes the sum of the square residuals (distances) from each data point to the regression line (termed *Y* estimate), hence the use of the term "least square" in connection with regression calculations used for calibration. Note that while Beer's Law is a rigorously derived model applying only to transmission spectroscopy, the same formulation is used for other spectroscopic methods of analysis. Unfortunately, we must forgo discussion of the pros and cons of this usage, as it is peripheral to the use of statistics, even though it is an important topic in its own right.

Beer's Law applies where refractive index, scattering, and specular reflection at "infinite–finite" numbers of surfaces all obey the Fresnel formulas. For reflectance spectroscopy, definitive, rigorous reflectance theory does not exist, as the convolution of an infinite number of integrals would be required to describe all of the combined light interaction effects at all surfaces, in all directions, at all angles, under varying conditions. Thus, Beer's Law is often used to illustrate the properties of reflectance spectroscopy for lack of an ideal mathematical model, and it serves adequately in a wide variety of cases.

So, how does one generate a calibration? Let us go back to basics and observe that (ideally) one can find a suitable wavelength for any given substance wherein absorption is proportional to concentration (Beer's Law). But as we noted, the real world is often non-ideal and absorbances deviate from Beer's Law at higher concentrations due most often to non-linearity of detection systems, scattering effects (which are wavelength dependent), and stray light-caused non-linearity. Note that spectra of compounds with absorbance bands narrower than the instrument resolution can demonstrate substantial non-linearities due to the equivalent of stray light within the bandpass characteristic of the instrument.

Even when one restricts measurements to lower concentrations where the relationship between change in absorbance and concentration is linear, the calibration line seldom passes through the origin. Accordingly, even for single components, an offset is most often observed. The use of derivative math pretreatment is often used to reduce the non-zero bias, yet derivatives do not remove multiplicative error and non-linearities. The offset occurs as a compensation to background interference and scattering. In a practical sense, most of the above considerations do not matter except in an academic sense. This is due to the fact that the practical tools used in spectroscopic methods are mostly based on empirical calibration methods where calibration techniques are well understood, although perhaps not always by the practicing chemist/spectroscopist who needs those tools. Calibration equations using multivariate techniques are used to compensate for the full variety of common variations found in "noisy" chemical values and imperfect instrumental measurements. This is why properly formulated calibration models work extremely well despite the imperfect world we find ourselves in.

If a set of samples is analyzed with high precision by some reference (standard) method so that the analyte concentrations are known, they can be used as a "teaching set" to generate an equation suitable for subsequent predictions. To comprise a good teaching set, the samples must evenly span the concentration range of interest. There is analytical "danger" in developing calibrations using sample sets with uneven constituent distributions, as in that case the mathematical calibration model will most closely fit the majority of samples in the calibration set. Therefore, if a calibration model is developed for a set of calibration samples weighted heavily toward one value, then that model will most closely fit samples at or near the mean concentration value, while giving poor results elsewhere. Conversely, when developing a mathematical model for an evenly distributed calibration set, the calibration model will have equal weight across the entire concentration range. Which situation is more desirable will depend upon the purpose for which the analytical results are to be used. Ideally, a properly developed calibration model will perform most accurately for samples at high and low parts of the concentration range when compared to calibrations developed using randomly selected samples, so that it does not matter where the "unknown" sample falls: an accurate result could be obtained anywhere.

The mathematical models that we presented above describe a straight line, and the presumption is that the data fall on that line. In fact, only the values predicted by the calibration equation fall on that line; for any given measurement of absorbance, the value specified by the line for the concentration corresponding to that absorbance value is the "predicted" (i.e., the spectroscopically determined) value for that sample. Note that the calibration points do not lie on that straight line, but are removed from a line by some distance (called a "residual"). The situation is depicted in Figure 32.1, which shows the relationship between the calibration line, the X-variable (absorbance, as measured by the instrument), the actual and predicted concentrations (which are both Y-variables, and are distinguished by calling them Y and $\hat{Y}$, respectively), and the residual, or the difference between the two Y-variables. Note how $\hat{Y}$ always falls on

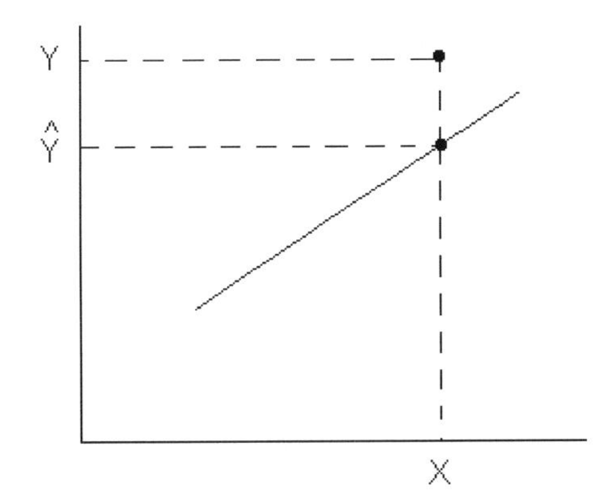

FIGURE 32.1. X is the value of the X-variable measured on a given sample, presumably absorbance in spectroscopic analysis. $\hat{Y}$ is the value of the constituent concentration that the calibration line (shown) determines (sometimes the term "predicts" is used) that the constituent concentration should be, if the model upon which that line is based is correct. Y is the actual value of the constituent concentration (or at least the value determined by some reference method) for that sample. The difference between them, ΔY, is called the *residual*. As discussed in the body of the text, the fact that the difference is measured vertically is of crucial importance.

the calibration line while Y does not necessarily do so. Another very important point to observe, although at first glance it seems trivial, is the direction in which the residual is measured: vertically, or parallel to the Y-axis. This point is crucial to the development and understanding of the effect of regression analysis on the calibration, as it is a consequence of the fundamental assumption of regression that there is no error in the X-variable. There are several consequences of this assumption (going from those of lesser to greater importance):

First, there is the question of units; if the direction of measurement of the error is not exactly vertical, the units of error must include a contribution of the units of the X-scale; keeping it vertical allows the error to be expressed solely in the units of the Y-scale (in our case, concentration units).

Second, statisticians have shown that if the error is measured in a direction other than the vertical (say, perpendicular to the calibration line), then the solution of the regression equations (which will be presented in the next chapter) is not necessarily unique; i.e., there may be more than one minimum in the sum of the squares of the errors, thus there may be more than one calibration line with the same "least square" error.

Third, and of most importance to us, the regression equations are developed under the assumption of no error in the X-variable. Thus, it is clear that the fundamental assumptions of regression analysis require that all the error of the analysis should indeed be, in spectroscopic parlance, in the reference analysis. This is fortunate, as this is often found to be the case. Allowing the error to include a component along the X-axis implies that this assumption fails to hold. In the real world, since the X-variable as well as the Y-variable is in fact the result of a measurement, it does in fact inevitably include some error also; successful calibration via regression analysis means that this error must be small enough to at least approximate the assumption of no error fairly well. A good deal of the difficulty and confusion encountered in using regression analysis to calibrate spectroscopic instruments is due to just this fact; that the assumption of no error in the X-variable is not sufficiently well approximated. We will eventually discuss means of detecting this situation and suggest some methods of correcting it.

With a mathematical treatment known as a "linear regression", one can find the "best" straight line through these real world points by minimizing the residuals. The resulting line is known as the "calibration" line, and its equation can be used to determine the concentration of unknown samples.

With more sophisticated algorithms, such as principal components regression (PCR) or partial least squares (PLS), we are regressing the sample scores rather than the optical data directly, but the above considerations still hold: all you must do is to replace the individual wavelength data in the equations we have presented with the corresponding values for the scores. A review of some of the calibration problems encountered using standard spectroscopic techniques with simple linear regression will be outlined in the next few chapters.

References

32-1. Draper, N., and Smith, H., *Applied Regression Analysis*, 3rd ed., Wiley, New York, 1998.
32-2. Kleinbaum, D.G., and Kupper, L.L., *Applied Regression Analysis and Other Multivariate Methods*, Doxbury Press, Boston, 1978.
32-3. Zar, J.H., *Biostatistical Analysis*, Prentice Hall, Englewood Cliffs, NJ, 1974.
32-4. Fogiel, M., "Regression and Correlation Analysis", *The Statistics Problem Solver, Research and Education Association*, New York, 1978, p. 665.
32-5. Mark, H., *Principles and Practice of Spectroscopic Calibration*, Wiley, New York, 1991.

33

Calibration: Linear Regression as a Statistical Technique

In any calibration method, the idea is to develop a mathematical model to define the relationship between some dependent variable (or variables) designated as Y, and other independent variables, designated as X. The simplest case of relating Y to X is where we use a single variable X to define Y. This procedure is defined as univariate regression. One of the basic problems in defining a mathematical model for X versus Y is to find the function that best describes the changes in Y with respect to changes in X. While sometimes a higher order function (polynomial, etc.) may be used to describe the relationship between X and Y, a straight-line model is the most common relationship. The trick is to find the relationship that best describes the change in Y with respect to X.

The simplest description of the X versus Y relationship is a straight line, and in our initial discussion, we will discuss only the straight-line case. When we draw a scatter plot of all X versus Y data, we see that some sort of shape can be described by the data points (Figure 33.1). With our scatter plot, we can take a basic guess as to whether a straight line or a curve will best describe the X,Y relationship. At this point in calibration development we are faced with three basic questions:

1. Is there some cause–effect relationship between X and Y? This non-statistical question is of utmost importance; if X and Y appear to be related but actually are not, our mathematical calibration model will have no meaning.

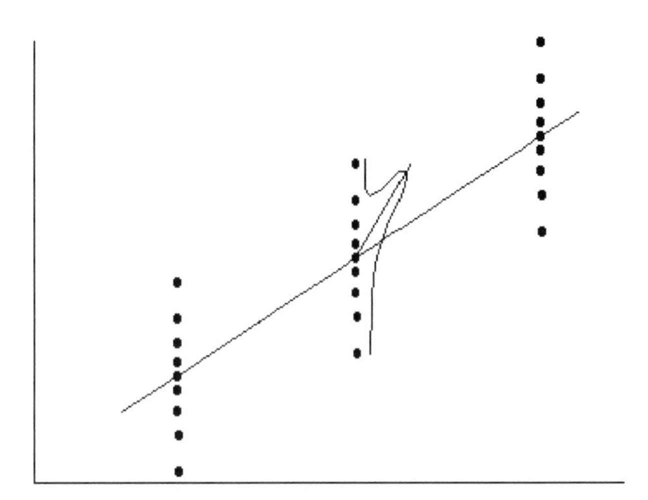

FIGURE 33.1. Scatter plot of X,Y data showing how errors are in Y only, and are Normally distributed around the calibration line.

2. If there is a relationship between X and Y, what is the best way to describe this relationship mathematically? Is a straight line best, or would a quadratic—or some other function—be better?
3. If we can decide on the best mathematical model to describe X versus Y, what is the best fitting curve or line to our data? In other words, how exactly does one describe the best line or curve that fits the data?

We begin by making an assumption about the best model. The assumption (or more properly the definition) is that the best-fitting model is the one with the smallest possible value of the sum of the squared errors. The reason for this goes back to something we demonstrated in Chapter 4: there we showed that the sample mean (the best estimate of the population mean) provides the smallest possible sum of squares of all possible numbers that the data could be compared to, and this was simultaneously the value *most likely* to be the population value under some very general assumptions. Similarly, since we compute the calibration from a sample of data, by imposing the criterion that it also be the least-squares estimator, we expect to arrive at a computed value for the calibration line that is also the best possible estimator of the population value for the calibration line—this is the reason that the least-square criterion is used. Statisticians call it the "maximum likelihood estimator", hearkening back to the origins of statistics in probability theory, but that is essentially just a more formal way of saying the same thing.

Also, if we choose a linear least-squares model, we can test it using statistics to determine how well the relationship between X and Y is described and how much variance or error is left over after we have done our best to describe Y using X. In order to test our model, we must first learn more about the statistics involved.

We begin by describing a line in mathematical terms, which looks something like this equation (remembering our previous discussion of the nomenclature of calibration equations in Chapter 32):

$$Y = B_0 + B_1 X \tag{33-1}$$

Nearly everyone has encountered this equation in a first algebra course and recognize that B is the designated symbol for the Y-intercept and B_1 is the slope term (see Equations 32-1–32-5). The slope term, as you will recall, is defined as a change in the Y value divided by a change in the X value for a straight line that fits the X and Y data. For us to be proper in our treatment of this subject, we must bring up at least two basic assumptions which are made in order for us to build a linear model between X and Y.

1. It is assumed that any errors that exist are independent of either the X or Y values.
2. It is assumed that for a fixed value of either X or Y that the corresponding variable, i.e., Y or X, respectively, have values that are random with a certain probability distribution. The distribution for both X and Y values is assumed to be Normal (Gaussian), as is the distribution of any errors. (It is this last assumption that allows the least-squares criterion to generate a maximum-likelihood estimator; if the errors are not Normally distributed then the maximum likelihood estimator is not necessarily the same as the least-squares estimator—recall our discussion of testing the distribution of data.) With only two variables we arbitrarily designate the one with the fixed values to be the X-variable. Excellent discussions of these assumptions are found in References (33-1–33-4).

Other assumptions have often been mentioned when describing the conditions for regression; these include:

3. Both the mean and the variance of Y depend upon the value of X.
4. The mean value for each distribution of Y is a straight-line function of X.
5. The variance of the X errors is zero, while the variance of the Y errors is identical for every value X. That is, the population variance of Y is exactly the same, for all values of X. When this is true, the variance of Y is called homoscedastic (X and Y have same/scatter); if this assumption is not true, the data are called heteroscedastic (X and Y have different/scatter).
6. The distribution of the errors in Y is Normal. This is shown in Figure 33.1.

The Least Squares or Best-Fit Line

The least-squares method is a procedure where the best-fitted straight line to a set of X,Y data points is determined according to the following criterion: This best-fit line occurs where the sum of squares for residuals (SS_{res}) is the least possible value when compared to any and all other possible lines fitted to that data. It is intuitively obvious that the smaller the distances from the data points in a plot of X,Y data to a line drawn through these data, the better does the line describe the data. In actuality, the distances from each data point to the line constitute unexplained variation, also termed unexplained error. In summation notation, the least-squares solution to the best-fit line to a set of X,Y data points is given by

$$\sum_{i=1}^{N} (Y_i - \hat{Y}_i)^2 = \sum_{i=1}^{N} (Y_i - B_0 - B_i X_i)^2 \tag{33-2}$$

where

B_0 = intercept for the best-fit line,
B_1 = slope for the best-fit line, the summation is taken over all the samples and this solution gives $\sum_{i=1}^{N} (Y_i - \hat{Y}_i)^2$ the minimum possible value.

The best least-squares fit line is then more precisely termed the minimum sum of squares. The minimum sum of squares is also referred to as the sum of squares for residuals or the sum of squares due to error.

In order to determine the minimum sum of squares, we must find the intercept (B_0) and the slope (B_1) that meet the criteria of minimum variance. In other words, we must calculate a slope and bias that provide the least amount of variance and can be both statistically determined and validated, so that both the bias and slope terms are known within a minimum confidence level.

How to Calculate the Best-Fit Line

We calculate the best-fit line intercept (B_0) by using

$$B_0 = \bar{Y} - B_1 \bar{X} \tag{33-3}$$

where B_1 is the slope as calculated below, and $\bar{X}$ and $\bar{Y}$ are the mean of X and Y values from all data points used.

To calculate the slope (B_1) for the best-fit line, we use

$$B_1 = \frac{\sum_{i=1}^{N} \{(X_i - \bar{X})(Y_i - \bar{Y})\}}{\sum_{i=1}^{N} (X_i - \bar{X})^2} \tag{33-4}$$

Then the overall line equation is given by

$$\hat{Y}_i = B_0 + B_1 X_i \tag{33-5}$$

where X_i is any x observation used to determine the best-fit line and $\hat{Y}_i$ is the estimated value of Y_i using the best-fit line with slope = B_1 and intercept = B_0.

How to Test the Best-Fit Line

Once we have calculated a best-fit line, it is a fine idea to determine the quality of the line fit to the data points used.

There are a variety of statistical tests used to evaluate goodness of fit for linear functions to any X,Y data. These include, but are not limited to, the 12 common statistics that we will come to in a later chapter.

These statistics are based on a simple but fundamental relationship among the various sums of squares that can be calculated from the regression. Recall these definitions from Chapter 32: Y is the constituent value for a given sample as determined be the reference laboratory method; $\hat{Y}$ is the constituent concentration for that sample as determined by the calibration model; $\bar{Y}$ is the mean value for all the values of that constituent (for the calibration data this is the same for both the reference values and the values from the calibration model). Using the analysis of variance techniques that we have discussed previously (see Chapters 9 and 26–29), the following relationship can be shown to exist among these quantities:

$$\sum(Y - \bar{Y})^2 = \sum(Y - \hat{Y})^2 + \sum(\hat{Y} - \bar{Y})^2 \qquad (33\text{-}6)$$

Exercise for the reader: prove this very important result shown in Equation 33-6 (use the discussions in the ANOVA chapters (26–29) as a model for how it is done—this is one reason we spent so much time on that topic—and do not be afraid to consult other statistical texts if necessary; by now you should be able to deal with them). Second exercise: set up the ANOVA table for this situation. What is the meaning of the F statistic for this case?

Equation 33-6 is an important key to understanding the meaning of the statistics that are used to evaluate a calibration. Let us look at what it tells us. First of all, the expression $\sum(Y_i - \bar{Y})^2$ is a constant; it depends only on the constituent values provided by the reference laboratory and does not depend in any way upon the calibration. The two terms on the right-hand side of Equation 33-6 show how this constant value is apportioned between the two quantities that are themselves summations, and are referred to as the sum of squares due to regression and the sum of squares due to error. We have already noted that by derivation of the regression calculations, the sum of squares due to error is the smallest possible value that it can possibly be for the given data. This being so, it is immediately clear that the sum of squares due to regression always attains the largest possible value that it can possibly have for the given data. All the global calibration statistics that we will study in the future come from this simple breakdown of the sums of squares.

The use of the sums of squares for residuals shown below:

$$\mathrm{SS}_{\mathrm{res}} = \sum_{i=1}^{N} (Y_i - \hat{Y}_i)^2 \qquad (33\text{-}7)$$

is useful for evaluating whether the calibration line is a good predictor of Y, given any value of X in the data range. If the SS_{res} is equal to zero, the calibration line is a perfect fit to the data. As the fit of the calibration line is worse, the SS_{res} increases due to the increased size of the residuals due to a lack of fit of the regression (calibration) line to the data. If there is a large SS_{res}, one of these facts would hold true:

(1) There is large variance within the data (i.e., σ^2 for the reference laboratory result is large).
(2) The assumption that a straight-line fit describes the relationship of change in Y with respect to X is false; a different model may be more appropriate, e.g., a non-linear fit.
(3) There is no significant relationship between X and Y, or there is no correlation between X and Y.
(4) There are unsuspected interferences in the samples, and being unsuspected, they were not included in the model.

Several statistics in later chapters will be discussed in terms of evaluating goodness of fit for calibration lines.

The ANOVA of Regression

The case that is, perhaps, more in line with our more general and widespread interest is the ANOVA for the case of a regression line, the question of which is posed above in conjunction with Equation 33-6. If you do not cheat and look at this answer before you try it for yourself, here is how it works:

From the diagram in Figure 33.2 it is clear that:

$$Y - \bar{Y} = (Y - \hat{Y}) + (\hat{Y} - \bar{Y}) \tag{33-8}$$

so that the sum of squares (over all the samples) is:

$$\sum(Y - \bar{Y})^2 = \sum((Y - \hat{Y}) + (\hat{Y} - \bar{Y}))^2 \tag{33-9}$$

Expanding the RHS:

$$\sum(Y - \bar{Y})^2 = \sum((Y - \hat{Y})^2 + 2(Y - \hat{Y})(\hat{Y} - \bar{Y}) + (\hat{Y} - \bar{Y})^2) \tag{33-10}$$

Expanding further and rearranging terms:

$$\sum(Y - \bar{Y})^2 = \sum((Y - \hat{Y})^2 + (\hat{Y} - \bar{Y})^2 + 2(Y\hat{Y} - \hat{Y}^2 - Y\bar{Y} - \hat{Y}\bar{Y})) \tag{33-11}$$

Collecting terms:

$$\sum(Y - \bar{Y})^2 = \sum((Y - \hat{Y})^2 + (\hat{Y} - \bar{Y})^2 + 2(\hat{Y}(\hat{Y} - \bar{Y}) + Y(\hat{Y} - \bar{Y}))) \tag{33-12}$$

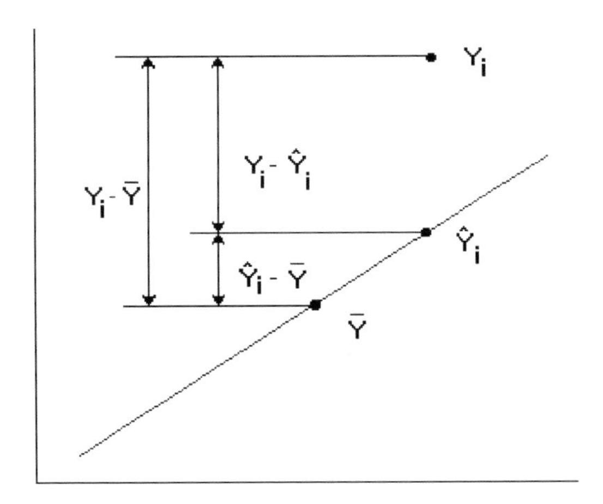

FIGURE 33.2. The diagram for the sum of squares for a calibration. $\bar{Y}$ represents the mean of all the data [for a calibration set the means of both the actual (reference laboratory) values and the predicted values (from the calibration model) are the same]. $\hat{Y}_i$ (read: Y-hat) is the value predicted from the calibration model for ith sample (and is therefore exactly on the calibration line). Y_i is the "actual" value (i.e., the reference laboratory value) of the ith sample from the reference laboratory. From the diagram the relationship between the three values, as described by Equation 33-8 in the text, is clear.

After distributing the summation, and noting that $\sum(\hat{Y} - \bar{Y})$ is identically zero, the final two terms drop out, leaving:

$$\sum(Y - \bar{Y})^2 = \sum(Y - \hat{Y})^2 + \sum(\hat{Y} - \bar{Y})^2 \qquad \text{(33-6(again))}$$

Thus, we arrive at the result that the total sums of squares is equal to the sums of squares of the two component parts; this is also explained in somewhat more detail in Reference (33-3, p. 18).

References

33-1. Kleinbaum, D.G., and Kupper, L.L., *Applied Regression Analysis and Other Multivariate Methods*, Doxbury Press, Boston, 1978.

33-2. Zar, J.H., *Biostatistical Analysis*, Prentice Hall, Englewood Cliffs, NJ, 1974.

33-3. Draper, N., and Smith, H., *Applied Regression Analysis*, 3rd ed., Wiley, New York, 1998.

33-4. Fogiel, M., "Regression and Correlation Analysis", *The Statistics Problem Solver, Research and Education Association*, New York, 1978, p. 665.

34

CALIBRATION: ERROR SOURCES IN CALIBRATION

In most calibration techniques one does not have to assume a linear relationship between the optical data and constituent concentration, as data transformations or pretreatments are used to linearize the optical data. The most common linearizing transforms include $\log(1/T)$ (for transmittance measurements) and either $\log(1/R)$ or the Kubelka–Munk function (for reflectance measurements) as math pretreatments. Calibration equations can be developed to compensate to some extent for the non-linear relationship between analyte concentrations and $\log(1/R)$ or Kubelka–Munk transformed data. PCR, PLS, and multilinear regression have been recommended to compensate for the non-linearity. Current evidence indicates that multilinear regression works better if the non-linearity is in the chemistry of the system (e.g., interactions between constituents), while the more sophisticated techniques (PCR, PLS) are superior if the non-linearity is in the optical data. If the matrix absorbs at different wavelengths than the analyte, Kubelka–Munk can prove to be a useful linearization method for optical data (34-1). If the matrix absorbs at the same wavelength as the analyte, $\log(1/R)$ will prove to be most useful to relate reflectance or transmission to concentration. When generating calibration equations using samples of known composition, the independent variable is represented by the optical readings ($-\log(R)$) at specific wavelengths, while the analyte concentration (as determined by manual laboratory technique) is the dependent, or Y-variable. Linear regression techniques allow for the selection of calibration spectral features that correlate (as a group) most closely to analyte concentration for a particular sample set. Once optimal wavelengths are selected, the spectrometer can be calibrated to predict unknown samples for the quantity of the desired analyte. Thus, regression analysis is used to develop the relationship between the spectral features and the chemical analyte (constituent) being investigated. Note that calibration equations will also contain terms at wavelengths that allow compensation for variations between sample aliquots and interferences within different samples due to background absorbance.

Questions often arise as to which mathematical treatments and instrument types perform optimally for a specific set of data. This is best addressed by saying that reasonable instrument and equation selection comprises only a small fraction of the variance or error attributable to calibration techniques. Actually, the greatest error sources in any calibration are generally due to reference laboratory error (stochastic error source), different aliquot error (non-homogeneity of sample—stochastic error source), and non-representative sampling in the learning set or calibration set population (undefined error). In general, it seems safe to say that there is more error variance due to the sample than due to the instrument/calibration combination. In Chapter 33 we used analysis of variance to demonstrate that the following relationship holds:

$$\sum_{i=1}^{N}(Y - \bar{Y})^2 = \sum_{i=1}^{N}(\hat{Y}_i - \bar{Y})^2 + \sum_{i=1}^{N}(Y_i - \hat{Y}_i)^2$$

Using a similar ANOVA technique, it is also possible to further break down the error contributions to the total variance, and thereby demonstrate that:

total variance due to error in analysis = sum of variances due to all error sources

Since the total variance is constant (in that it depends only on the reference laboratory values for the calibration sample set), it is clear that the largest contributions to calibration error can be minimized only by reducing the major contributors to variance. Reducing or even eliminating entirely the error due to a minor source will make no appreciable difference to the total error term. The major error sources are shown in Table 34-1 and some selected ones discussed in the following text:

(1) Population error. Population sampling error can be minimized by collecting extremely comprehensive data sets and then reducing them via subset selection algorithms. These techniques, which have been described (34-2,34-3), allow maximum variation in a calibration set with minimum laboratory effort and are implemented in software available from instrument manufacturers.

TABLE 34-1
Major calibration error sources.

* Sample non-homogeneity
* Laboratory error
* Physical variation in sample
* Chemical variation in sample with time
* Population sampling error
* Non-Beer's Law relationship (non-linearity)
* Spectroscopy does not equal wet chemistry
* Instrument noise
* Integrated circuit problem
* Optical polarization
* Sample presentation extremely variable
* Calibration modeling incorrect
* Poor calibration transfer
* Outlier samples with calibration set
* Transcription errors

The question often asked is, "how many samples should be used for calibration?". The correct answer is, "enough to assure that all the relevant sources of variation are sampled and thereby included in the calibration set"; unfortunately, this answer is not very useful in practice. In general, the more samples that are included, the greater the probability that all the necessary variances are indeed sampled, even though they might be unknown. A good rule of thumb is 10 samples, plus 10 for each constituent to be calibrated.

(2) Laboratory error. This source of error can be substantially reduced by performing an in-house audit of procedures, equipment, and personnel, paying particular attention to sample presentation, drying biases, and random moisture losses upon grinding (34-4). The contribution of this error source to the total variance can be assessed using the ANOVA techniques described in Chapter 28.

(3) Sample presentation error. Presentation variation can be accommodated by compression (averaging) multiple sample aliquots, and by generating a calibration equation on the compressed data and then by predicting similar averages of unknown samples (34-5). Spinning or rotating sample cups also can reduce this error. Continuous rotation while collecting the optical data introduces other errors into the calibration; a preferred method is to rotate the sample, then stop the rotation, with a different segment of the sample in the active optical region, for data collection; this is equivalent to automatic resampling of different aliquots. The concept is to produce as large a population of measurements as possible noting that the mean of this set of measurements more closely approximates a less "noisy" measurement value. The combination of these methods can reduce prediction errors by as much as 70% (relative) for predicted values as compared with less careful sample collection and

presentation. In the absence of an automatic means of resampling, manually repacking a sample and averaging multiple repacks has also been successfully used to reduce this error source (see Reference (34-5)).

(4) Wavelength selection error. The basic question here is: how can we determine a confidence interval for wavelength selection? Alternatively, how do we represent the possibility of disparate sets of wavelengths that might be chosen?

The actual choice of wavelengths would be determined by the propagation of noise through the inversion of the variance–covariance matrix. This depends on the actual noise of the data, and the condition number of the matrix to be inverted.

There are several possible conditions under which multiple regression calibration calculations can be performed.

(A) No noise on either the dependent variable or any of the independent variables. In this case, regression calculations are unnecessary, it will suffice to solve simultaneous equations.
(B) Noise on the dependent variable only. This is the case for which the regression calculations are defined, by the standard considerations.
(C) Noise on the independent variables only. This case is unsolved, except for an interesting special case. Specifically, in the case of simple regression (i.e., only one independent variable) the variables can be interchanged, calculations performed on the data where the variable that is now the "independent" variable is noise-free, and the desired slope is then reciprocal of the slope calculated from the reversed data.
(D) Noise on both types of variables. This case has been studied, but is also unsolved. It is well known that the calculated slope of the regression is a biased estimator of the beta. For simple regression, the results are well known, and the amount of bias is calculable if the noise of the independent variable is known. For multiple regression, the results are much more complicated and not well determined.
(E) The variable used to measure the analyte may exhibit curvature with (i.e., not be a linear function of) the amount of analyte (i.e., the dependent variable). This will show up in the residuals.
(F) A variable used to measure one of the interfering constituents may exhibit non-linearities. This case is different than E, because this affects the degree of correction for the interference, and may not show up in the residual plot.

(5) Outlier prediction is important during the calibration modeling and monitoring phases. True spectral outliers are considered to be samples whose spectral characteristics are not represented within a specified sample set. Outliers are not considered to be part of the group that is designated to be used as a calibration set. The criterion often given representing outlier detection is a sample

spectrum with a distance of greater than three Mahalanobis distances from the centroid of the data. Another quick definition is a sample where the absolute residual value is greater than three to four standard deviations from the mean residual value (34-6). We refer the reader to standard multivariate or regression texts, which more extensively describe these selection criteria; the last word on this topic has got to be Reference (34-7).

In a practical sense, outliers are those samples that have unique character so as to make them recognizably (statistically) different from a designated sample population. Evaluation criteria for selecting outliers are often subjective; therefore, there is a requirement that some expertise in multivariate methods be employed prior to discarding any samples from a calibration set.

There are a number of ways in which an "outlier" might differ from the target population. First, it might be a perfectly normal sample, but the reference laboratory value for the constituent may have a gross error. Unfortunately, this is all too common a situation. This type of error will cause the sample to appear as an outlier on a residual plot, and, indeed, it should be deleted from the sample set. There is apparently a good deal of misinformation prevalent about this deletion process: it seems to be widely believed that the reason to delete the sample is to get the bad value out of the data set, so that the calibration will look better. In fact, the reason for deletion of these types of outliers is not because the outlier itself is bad, but because it will tend to draw the calibration line away from its proper position through the good data, thereby causing bad results to be generated from the rest of the perfectly good samples in the calibration set.

A second major cause of outliers is samples that are, in fact, not part of the target population. This may be because they contain extreme values of one or another constituent, or an unusual combination of constituents (even when each constituent is within its normal range individually), or there may be some generic difference between that sample and the rest of the samples in the calibration set. Strictly speaking, effects such as poor sample preparation or a bad pack of a solid sample into the cell also fall into this category, if all the other samples in the set were well prepared and properly packed. If the sample is, indeed, not part of the target population, then the purpose of the calibration should be reassessed, and perhaps the target population should be redefined. If the target population is to include the "outlier", then more samples of that type must be collected and included in the calibration. If the outlier is to be rejected, then a decision must be made as to how to deal with such samples when they show up in the future.

The third major cause of outliers is bad spectral data. This could be due to instrumental effects (noise, drift, etc.), sample inhomogeneity, non-linearities, hardware breakdowns, etc. Strictly speaking, the proper way to deal with these causes is to find and correct the problem. In practice, these are also often dealt with by sample deletion. Sample deletion is OK in this case, as long as not too many samples are deleted for this cause; however, the real problem with simply deleting samples from the calibration set is that it does not cure the problem, and if/when the problem occurs during routine analysis, there may be no indication, and an incorrect analytical value for the routine sample will be generated.

References

34-1. Olinger, J.M., and Griffiths, P.R., *Analytical Chemistry*, 60, (1985), 2427.

34-2. Honigs, D.E., Hieftje, G.M., Mark, H.L., and Hirschfeld, T.B., *Analytical Chemistry*, 57: 12, (1985), 2299–2303.

34-3. Shenk, J., "Equation Selection", Marten, G.C., Barton, F.E., Barton, F.E., and Shenk, J.S. eds. *Near Infrared Reflectance Spectroscopy (NIRS); Analysis of Forage Quality*, USDA Agric. Handbook #645, National Technical Information Service, Springfield, VA, 1986, pp. 26–27.

34-4. Youden, W.J., and Steiner, E.H., *Statistical Manual of the AOAC*, 1st ed., Association of Official Analytical Chemists, Washington, DC, 1975.

34-5. Mark, H., and Workman, J., Analytical Chemistry, 58: 7, (1986), 1454–1459.

34-6. Draper, N., and Smith, H., *Applied Regression Analysis*, 3rd ed., Wiley, New York, 1998.

34-7. Barnett, V., and Lewis, T., *Outliers in Statistical Data*, Wiley, New York, 1978.

35

Calibration: Selecting the Calibration Samples

The selection or preparation of a set of calibration standards involves the following important considerations. The analyst must collect samples that include the complete range of component concentration as evenly distributed as possible. Random sample selection will cause the mathematical model to be most closely fit to the middle concentration samples, while the few samples at extremely high or low concentration levels will influence the slope and intercept inordinately. An even concentration distribution will allow the model to minimize the residuals at the extremes and at the center with relatively equal weighting. Note: In the past, weighted regression techniques have not received widespread use in NIR; thus, the general case for the most commonly used regression algorithms is discussed here.

Besides sample distribution, another critical consideration in selecting/preparing standards is uniform matrix distribution. What is meant here is that the variance in background composition can bring about substantial changes in spectra, such as band broadening and peak position shifts. Background characteristics must then be carefully considered when composing the standard calibration set. For example, if the polarity of the selected samples changes with the concentration of various matrix components (e.g., pH), and a polarity change brings about band shifts in the spectra, then in order to develop a mathematical model for this application, a range of pH variations must be included in the calibration standards. The pH must vary throughout the range of pH's that might be found in the future samples to be run; furthermore, it is important that the pH vary independently of the other constituents in the sample set. A calibration model is then developed representing the types of samples that will be analyzed in the general analysis procedure, including the variations in pH. Another example often found with solid samples is the effect of moisture content within a solid or powdered sample. The presence or absence of water in a sample will, of course, influence the extent of hydrogen bonding within the sample. Hydrogen bonding will affect both band position and width. If a mathematical model is developed on standards that include a wide range of the component of interest but only a small range in moisture, the calibration model will only be useful for samples with a narrow moisture range such as was represented by the standards. Samples with a moisture content outside the range of the calibration samples will not give correct results, just as in the case of samples whose constituent values lie outside the calibration range. Each calibration problem represents something slightly different, yet successful calibration can be performed by paying attention to these finer details.

Other common error sources that are included are technician error in sample preparation, temperature differences in standards or instrument while taking data, calibration standard instability, instrument noise and drift, changes in instrument wavelength setting, non-linearity, stray light effects, particle size differences, color differences with concentration, differences in the treatment of the sample (e.g., different dye colors in textiles and fabrics), solvent interaction differences with change in concentration, or the reference method does not measure the same component as a spectroscopic method.

An example of the last problem described above would typically be found in measuring the percent composition of protein in a biological solid or liquid. A thermal analysis or Kjeldahl procedure might be used as a reference method to provide calibration values. Both thermal analysis and Kjeldahl procedures produce measurements of total reduced nitrogen content. A near-infrared spectral measurement determines information on peptide bonds directly, but not on reduced nitrogen alone. In this type of application, the spectroscopic data never will perfectly agree with the reference data. A review of such difficulties, and other, similar ones in chemical analysis has recently appeared (35-1).

When creating a calibration set using volumetric or gravimetric techniques, there are several important items to consider. Often, secondary characteristics such as the refractive index of liquids change with concentration; particularly problematic are high concentration standards. This problem can be overcome by operating in narrow ranges that approximate linearity; a term sometimes called "range splitting". The measured absorptivity of a mixture is dependent upon the bandpass of an instrument. By leaving the instrument settings of bandpass, scan speed, and response fixed during analysis, this problem can be minimized.

The narrower the absorption band of the sample as compared to the bandwidth of the instrument, the more deviation from Beer's Law occurs. This is due to the fact that a smaller percentage of the total energy that passes through or into the sample is attenuated by the absorbing substance. This is because if the absorption band is narrow, then a greater proportion of light that enters the sample bypasses the region of absorbance, and therefore it is not absorbed. Thus, a greater percentage of the energy in the incident beam is not affected by the absorbing band and reaches the detector. Broad bands measured with relatively narrow bandpass instruments do not suffer this problem.

Synthetic samples are generally not considered useful for developing calibration models due to several of the above considerations. A rule of thumb for synthetic samples dictates that if there is any spectroscopically observable interaction among the components in the mixture, it will be impractical to produce a standard calibration set. If the only option for the analyst is to produce synthetic standard samples, then rules of mixture design must be followed. In composing standards, it has been an unofficial rule of thumb that 10 or more samples per term (wavelength or independent variable included in the calibration equation) should be used in the mathematical model or equation. Actually, as many samples as required to represent variability of the samples to be measured is a technical requirement. Experimental design can assist the user in composing a calibration set when the total components or variables within the sample have been estimated. This can be done using principal component analysis (35-2) or by surveying the statistics literature or the applications literature for the experience of previous researchers. For powdered samples, particle size and moisture (O–H stretch) often comprise the first two major components. Mixture design concepts can be useful in understanding the calibration model concept.

Selecting a sample set for calibration involves logic and understanding similar to that used by pollsters to predict elections. The calibration experiment, however, involves variables that are less confusing than election polls and more stable, because they are based upon physical principles; therefore, much greater confidence is placed upon these analytical calibrations. For example, there is generally a substantial *a priori* knowledge of the chemistry involved in the analytical samples prior to the calibration. Other variables, such as temperature, instrumental variations, and the like are, in principle at least, well understood. In the election situation, war, scandal, economic uncertainty, or revolution can create unpredictable results. Yet, polling science is useful in predicting results from large open populations. Pollsters compensate for the interfering events by correct design of their experiment and proper selection of their sample. How much more confidence we should have, then, that the resulting calibration will accurately predict unknown samples if we could ensure that our calibration set is properly selected.

Calibration sets must not only uniformly cover an entire constituent range, they must also be composed of a uniformly distributed set of sample types. Ideal calibration sets are composed of a number of samples equal to or greater than 10–15 samples per analytical term. These samples ideally have widely varying composition evenly distributed across the calibration range. The ideal sample set will also not exhibit intercorrelation between moisture, particle size, or any other characteristic with any of the components of interest. An exception to this intercorrelation rule is allowed in obvious cases, such as total solids determination or some such application where moisture is an indirect measurement of the total solids. However, one must be able to rely on the intercorrelation: if, for example, another volatile component is present in some samples but not in the ones used for calibration, then the "total solids" will be affected by this other material, leading to increased error.

In another well-documented case, the hardness of wheat is directly related to the particle size distribution of the ground wheat due to the grinder/wheat interaction: the harder wheats tend to be ground to larger average particle sizes than softer wheats.

The variance of the sample set used to produce near-infrared calibration equations determines both robustness and accuracy for a particular application. A calibration learning (teaching) set, which includes a wide variation of sample types and a large constituent range, will allow a calibration model where a wider range of materials may be analyzed, but sometimes with a concomitant loss in accuracy. If the calibration learning set has a small variance in sample type and a narrow constituent range, the accuracy for well-analyzed samples within the range is increased, but fewer unusual samples can be tolerated or analyzed with confidence, using this approach. Thus, for quality control procedures, one may wish to have both calibrations available. The robust calibration is used to analyze those samples that appear as "outliers" for the dedicated calibration, while the dedicated calibration is used to measure accurately the constituent values of "normal" samples. Outlier detection techniques can be used to predict the "uniqueness" of a sample using the H statistic, also known as Mahalanobis distance.

References

35-1. Rogers, L.B., *Analytical Chemistry*, 62: 13, (1990), 703A–711A.
35-2. Cowe, I., and McNichol, J.W., *Applied Spectroscopy*, 39, (1985), 257.

36

CALIBRATION: DEVELOPING THE CALIBRATION MODEL

Table 36-1 demonstrates the common methods for reducing large arrays of data (spectra, in our case) to fewer data points. Note that first- through fourth-order functions are used with either zero or natural y intercepts. The method of least squares is used to obtain these calibration equations. When a small number of standards are used or when the relationship between the optical data and concentration is not known, a segment fit can be used. The segment fit assumes nothing about the data other than the facts that each datum has the same significance in the calibration and that predicted values are based only on series of lines that connect the data points. A segment fit produces no goodness-of-fit statistics due to the fact that it always forms a "perfect" fit to the data. It can be useful in single-component, rapid-analysis, or feasibility studies. A complete univariate software package will include segment fitting as a convenience for analysts performing "quick and dirty" experiments, but this type of calibration is not particularly useful for serious work. Multicomponent methods, also termed multivariate techniques, are used when a single wavelength will not predict concentration due to band overlap or when more than one component is to be measured simultaneously.

Simple multicomponent measurements are sometimes used when the spectra of two or more chromophores within a mixture are found to be overlapped across the entire spectral region of interest. When performing multivariate analysis on

TABLE 36-1
A description of independent variable terms (math pretreatments or data decomposition techniques).

Name of independent variable terms	Equivalence in $-\log(R)$ terms
$-\log(R)$	1
First difference ("derivative")	2
Second difference ("derivative")	3
Higher order differences ("derivatives")	≥ 4
Difference ratios	≥ 4
Fourier coefficients	1 to full spectrum
Principal components scores	1 to full spectrum
Factor analysis scores	1 to full spectrum
Curve fitting coefficients	1 to full spectrum
Partial least squares scores	1 to full spectrum

such mixtures, the analyst must select wavelengths where the molar absorptivities for the components are most different. With complex matrices, it is impossible to know if an ideal set of wavelengths has been selected for a calibration equation. Often, when new instrumental effects are indicated and when new calibration sets are selected, new wavelengths are chosen for the same application. It has been somewhat dissatisfying to be caught in a trap of juggling wavelengths always hoping to find the best combinations for a particular analytical task. In multilinear regression, the analyst assumes that adding new or different wavelengths will provide a calibration equation compensating for noise, interferences, non-linearity, background scattering, and other "nasty" deviations from Beer's Law; sometimes this is true and sometimes not (36-1–36-6). It is particularly frustrating when the realization dawns that some of the effects seen in the data, such as the variations due to such physical phenomena as particle size, clearly affect the entire spectrum in a systematic manner, so that it is clear that at data from at least one wavelength are needed to accommodate this extra variable in the data. Yet it is equally clear that a phenomenon such as "particle size" simply has no absorbance band to key in on and use as a guide for wavelength selection.

Principal components regression (PCR) has been applied in a variety of calibration techniques with reported success. This technique has the benefit of not requiring wavelength selection. The analyst must decide how many components (or eigenvectors) to use in the calibration equation. Since the eigenvectors are exactly orthogonal, the choice is considerably simplified compared to using data at individual wavelengths. Partial least squares (PLS) is another multivariate technique that seems to show promise in calibrations where samples are most complex and where only weak relationships exist between analyte and spectral features. PLS has wide-ranging applications in a variety of spectroscopic and scientific applications. PLS also shows promise in color modeling. Table 36-1 illustrates the most common mathematical pretreatments and their equivalence in log(1/*R*) terms. These methods will be discussed in greater detail in later chapters.

Validating the Calibration Model

Table 36-2 lists the statistics used to determine equation suitability for linear and multilinear least squares. The following symbols can be used to describe statistical terms used to validate equation models:

R^2 = coefficient of multiple determination;

$\sum$ = capital sigma represents summation, or sum of all values following it;

N = Notation for total number of samples used for a computation;

Y_i = a single *Y*-value for the *i*th sample. In real cases it represents all the actual *Y*-values for the 1st through *N*th data points;

$\hat{Y}_i$ = the symbol for the estimated *Y*-value, given a regression line. Therefore, for any given set of *X*-values, for an *i*th sample there is a corresponding Y_i or estimated value based on the X_i value. It can also be stated as an analytical value derived by the calibration equation;

$\bar{Y}$ = the mean *Y*-value for all samples where *Y* is some function of *X* as $Y = f(X)$;

K = the number of wavelengths used in an equation;

R = the simple correlation coefficient for a linear regression for any set of data points; this is equal to the square root of the coefficient of multiple determination. Note: a lot of people get hung up over the question of whether an upper case or lower case symbol (or sometimes a different label) represents a particular statistic. While precision in terminology is laudable, different authors have used different ones, so the key here is not to worry about the symbol used for a given

TABLE 36-2
Statistics used to evaluate equation suitability.

1. Coefficient of multiple determination
2. *F* test statistic for the regression
3. Confidence limits for predicted values
4. Student's *t*-value for a regression
5. Studentized *t* test for the residual between the calibration estimated value and the actual value for any sample
6. Partial *F* or *t*-squared test for a regression coefficient
7. Standard error of estimate
8. Standard error of prediction
9. Standard error of cross validation
10. The bias-corrected standard error
11. Standard deviation of difference
12. Standard error of laboratory

statistic, but make sure you know what the discussion is about—then the meaning will be clear from the context;

B_0 = the y-intercept value for any calibration function fit to X,Y data. For bias corrected standard error calculations, the bias is equal to the difference between the average primary wet chemical analytical values and the NIR predicted values;

B_i = regression coefficient at one wavelength.

The first 11 statistics listed in Table 36-2 for validating calibration equations can all be described by using the following five basic statistical quantities:

1. Sum of squares for regression (SS_{regr}):

$$SS_{regr} = SS_{tot} - SS_{res} = \sum_{i=1}^{N} (\hat{Y}_i - \bar{Y})^2$$

   Review our description of these various sums of squares in Chapter 33.

2. Sum of squares for residuals (SS_{res}):

$$SS_{res} = \sum_{i=1}^{N} (Y_i - \hat{Y}_i)^2$$

3. Mean square for regression (MS_{regr}):

$$MS_{regr} = \frac{SS_{tot} - SS_{regr}}{\text{d.f. for regression}} = \frac{\sum_{i=1}^{N} (\hat{Y} - \bar{Y})^2}{K + 1}$$

4. Mean square for residuals (MS_{res}):

$$MS_{res} = \frac{SS_{res}}{\text{d.f. for residuals}} = \frac{\sum_{i=1}^{N} (Y_i - \hat{Y})^2}{N - K - 1}$$

5. Total sum of squares (SS_{tot}):

$$SS_{tot} = \sum_{i=1}^{N} (Y_i - \bar{Y})^2$$

We will begin a detailed consideration of the statistics in Table 36-2 in the next chapter.

References

36-1. Sustek, J., *Analytical Chemistry*, 46, (1974), 1676.

36-2. Honigs, D.E., Freelin, J.M., Hieftje, G.M., and Hirschfeld, T.B., *Applied Spectroscopy*, 37, (1983), 491.

36-3. Maris, M.A., Brown, C.W., and Lavery, D.S., *Analytical Chemistry*, 55, (1983), 1694.
36-4. Kisner, M.J., Brown, C.W., and Kavornos, G.J., *Analytical Chemistry*, 55, (1983), 1703.
36-5. Otto, M., and Wegscheider, W., *Analytical Chemistry*, 57, (1985), 63.
36-6. Frans, S.D., and Harris, J.M., *Analytical Chemistry*, 57, (1985), 2580.

37

CALIBRATION: AUXILIARY STATISTICS FOR THE CALIBRATION MODEL

In this chapter, we will discuss some of the statistics that are commonly used in conjunction with the multivariate calibration techniques used for spectroscopic analysis. We say "some" because no listing could possibly be complete due to the fact that each statistician has his own favorite set of statistics, and while some are in more widespread use than others, there are a good number of them that have their own particular adherents. Here we describe the ones that we feel are both useful and fairly common in spectroscopic calibration work, so that anyone reading the literature is likely to come across them. Since we know that our list cannot be all-inclusive, our best advice in case you run across one that is not in our list is simply to follow up on finding out the definition (the mathematical definition, that is) of the term, and try to relate it to the ones that you already know.

We will try to present our list in a consistent format and to list as many of the pertinent aspects of these statistics as we can. Our format will be:

(1) The name of the statistic.
(2) The abbreviation (or symbol) for that statistic; we also include synonyms.
(3) The calculational formula in summation notation.
(4) Other computational forms; often these allow shortcut methods of calculation, or show the relationship to other statistics. Recall from

Chapter 36 that most calibration statistics are derivable from the various sums of squares obtained from the regression.

(5) Comments: any useful information that we feel will help.

Note: all summations are taken over N, the number of samples in the set, unless otherwise specified. This allows us to simplify the notation by leaving the limits of summation off the summation signs where convenient.

1. *Statistic:* Coefficient of multiple determination
Abbreviations: R^2 or r^2
Summation notation:

$$R^2 = 1 - \frac{\sum(Y_i - \hat{Y}_i)^2/(N-K-1)}{\sum(Y_i - \bar{Y})^2/(N-1)} = \frac{\sum(\hat{Y}_i - \bar{Y})^2}{\sum(Y_i - \bar{Y})^2}$$

Computational formula:

$$R^2 = 1 - \left(\frac{\text{SEC}}{\text{S.D.}_{\text{range}}}\right)^2 = \frac{\text{SS}_{\text{tot}} - \text{SS}_{\text{res}}}{\text{SS}_{\text{tot}}}$$

Comments: Also termed the R-squared statistic, or total explained variation. This statistic allows us to determine the amount of variation in the data that is adequately modeled by the calibration equation as a fraction of the total variation. Thus, $R^2 = 1.00$ indicates the calibration equation models 100% of the variation within the data. An $R^2 = 0.50$ indicates that 50% of the variation (in the differences between the actual values for the data points and the predicted or estimated values for these points) is explained by the calibration equation (mathematical model), and 50% is not explained. R-squared values approaching 1.0 are attempted when developing calibrations. The highest value of R^2 attainable from a given sample set can be estimated using this calculation:

$$R^2_{\text{MAX}} = 1 - \left(\frac{\text{SEL}}{\text{S.D.}_{\text{range}}}\right)^2$$

where SEL = the standard deviation of the errors of the reference method and $\text{S.D.}_{\text{range}}$ = the standard deviation of all the reference values of the constituent in the calibration dataset.

2. *Statistic:* F test statistic for the regression [Note: do not confuse this with Statistic 5 (Studentized t for the residual) or with Statistic 6 (partial F for a regression coefficient)—see discussion for Statistic 5.]

Abbreviations: F, t^2
Summation notation:

$$F = \frac{\sum(\hat{Y}_i - \bar{Y})^2/K}{\sum(Y_i - \hat{Y})^2/(N - K - 1)}$$

Computational formula:

$$F = \frac{R^2/(N - K - 1)}{(1 - R^2)/K} = \frac{\mathrm{MS}_{\mathrm{regr}}}{\mathrm{MS}_{\mathrm{resi}}}$$

Comments: Also termed F from regression, or t^2 (for a simple, one-wavelength regression only. For multiple regression situations, the relationship to the t statistic is more complicated). This is the F statistic that we computed as the result of an analysis of variance for the calibration in Chapter 33. (Exercise for the reader: go back and check this out, and recall the steps involved in reaching this point.) F increases as the equation begins to model, or fit, more of the variation within the data. Thus, the F statistic increases as the coefficient of multiple determination increases, and does so very rapidly as R^2 approaches unity. Also, as the computational formula shows, for a given value of R^2, the F value increases as the number of samples increases, and as the number of wavelengths used within the regression equation decreases. Thus, deleting an unimportant wavelength from an equation will cause the F for regression to increase, since K in the denominator of the computational formula will decrease while R^2 remains essentially constant. However, if a wavelength important to the calibration is deleted, then the term involving R^2 will decrease faster than the term involving K causes the expression to increase; therefore, F will also decrease. Thus, F is a sensitive indicator of the overall "goodness" of the calibration.

The F statistic can also be useful in recognizing suspected outliers within a calibration sample set; if the F value decreases when a sample is deleted, the sample was likely not an outlier. This situation is the result of the sample not affecting the overall fit of the calibration line to the data while at the same time decreasing the number of samples (N). Conversely, if deleting a single sample increases the overall F for regression, the sample is considered a suspected outlier. F is defined as the mean square for regression divided by the mean square for residual (see statistical terms in the table in Chapter 36).

3. *Statistic:* Confidence intervals for predicted values
Abbreviation(s): $\hat{Y} \pm$ C.L. Also termed confidence bands, confidence intervals for a calibration line, or 95% confidence limits for a calibration line.

Summation notation:

$$\text{(Confidence interval)} = \hat{Y} \pm (2F)^{1/2}\left\{\frac{\sum(Y_i - \hat{Y}_i)^2}{N-K-1}\left[\frac{1}{N} + \frac{(X_i - \bar{X})^2}{\sum(X_i - \bar{X})^2}\right]\right\}^{1/2}$$

where F is $F(1 - \alpha, 2, N - K - 1)$ from a table of the F distribution.
Computational formula:

$$\text{(Confidence interval)} = \hat{Y} \pm (2F)^{1/2}(\text{SEC})\left[\frac{1}{N} + \frac{(X_i - \bar{X})^2}{\sum(X_i - \bar{X})^2}\right]^{1/2}$$

Comments: We note that the confidence interval for any statistic is equal to the statistic $\pm t$ (or the square root of $2F$) times the standard error of the statistic. In calibration, the confidence limit allows us to estimate the precision of the analytical method for samples of differing actual concentrations. Often confidence limits are shown diagrammatically as being bounded by the two branches of a hyperbola as shown in Figure 37.1. This is due to the fact that the slope of the line also has a confidence interval; this manifests itself by the fact that the line far away from $\bar{X}$ moves around more (is more variable) than the line near $\bar{X}$. This relationship can be calculated for most X,Y situations where $Y = f(x)$.

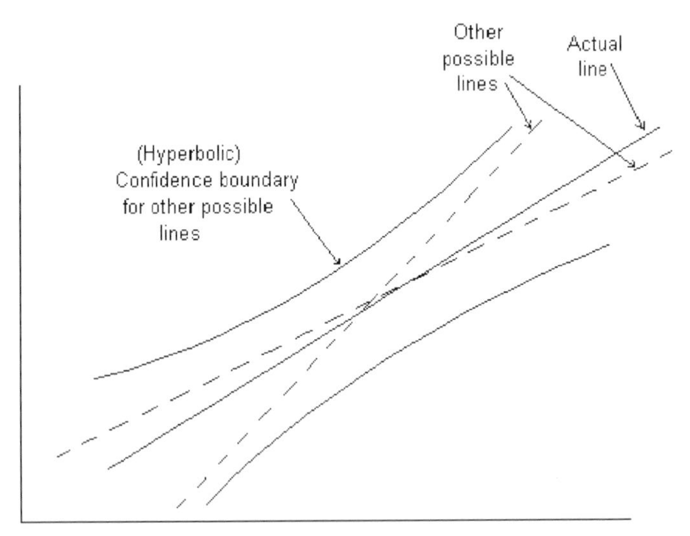

FIGURE. 37.1. Different sets of samples of the same type will have different sets of errors (by random chance alone). Therefore, calibrations on such sets will result in different calibration lines. The probability structure of the possible calibration lines is bounded by confidence hyperbolas, one of which is shown.

4. *Statistic:* Student's t value (for a regression)
Abbreviation(s): t, $F^{1/2}$, t test for regression ($F^{1/2}$ is for a simple, one-wavelength regression only. For multiple regression situations, the relationship between F and the t statistic is more complicated)
Summation notation: (see F statistic for the regression)
Computational formula:

$$t = \frac{R(N - K - 1)^{1/2}}{(1 - R^2)^{1/2}} = F^{1/2} = \left\{ \frac{\text{MS}_{\text{regr}}}{\text{MS}_{\text{resi}}} \right\}^{1/2}$$

Comments: This statistic is equivalent to the F statistic in the determination of the correlation between X and Y data. It can be used to determine whether there is a true correlation between a spectroscopically measured value and the primary chemical analysis for that sample. It is used to test the hypothesis that the correlation really exists and has not happened only by chance. A large t value (generally greater than 10) indicates a real (statistically significant) correlation between X and Y; this can be tested against the standard statistical tables for the percentage points of the t distribution.

A point sometimes missed is that doing a hypothesis test and finding a statistically significant value for t (or for any other calibration statistic, for that matter) shows only that the evidence for association among the variables involved is too strong for it to have happened by chance alone, so that the association is real. It does not tell whether the calibration for which the statistics are calculated is able to serve any useful purpose, i.e., is "good enough" for the desired analysis.

5. *Statistic:* Studentized t test for the residual or the difference between the calibration estimated value and the actual value for any sample. Note: do not confuse this t value with the one previously discussed (Statistic 4) or another one we will discuss shortly (Statistic 6). The result of dividing any statistic by its standard error has a t distribution (for the associated degrees of freedom). Thus, the important distinction to make here is: what is the base variable that is involved? In the previous statistic the variables involved were the mean squares for the regression and for error. In this case it is the error in the reading of an individual sample. In the case of Statistic 6 the base variable is the calibration coefficient.
Abbreviations: t, $F^{1/2}$
Summation notation: See discussion of SEC.
Computational formula:

$$t = \frac{\text{Residual}}{\text{SEC}} = \frac{Y_i - \hat{Y}_i}{\text{SEC}}$$

Comments: This test creates evaluation criteria for assessing the variation between an NIR value and its primary chemical value. To avoid having to go to the t tables, a rule of thumb is that t values of greater than 2.5 are considered significant and those spectroscopic analyses having such large or larger t values may possibly be outliers. Replicate primary reference analysis should be performed on such samples. If the wet chemistry is proven correct, the sample may be a spectral outlier. Most often, high t test values here indicate poor laboratory results or a problem with sample presentation, such as grinding or packing of the sample cup.

6. *Statistic:* Partial F or t^2 test for a regression coefficient (not to be confused with Statistic 2 or Statistic 5—see discussion for Statistic 5).
Abbreviation(s): F, t^2
Summation notation:

$$t = \frac{b_i}{\mathrm{SE}(b_i)}$$

Computational formula:

$$F = \frac{\mathrm{SS}_{\mathrm{res}_{\text{all variables except the one under test}}} - \mathrm{SS}_{\mathrm{res}_{\text{all variables}}}}{\mathrm{MS}_{\mathrm{res}_{\text{all variables}}}}$$

Comments: For use in multivariate calibration modeling. This test indicates if the deletion of a particular wavelength (independent variable) and its corresponding regression coefficient (multiplier) cause any significant change to an equation's ability to model the data (including the remaining unexplained variation). Small F or t values indicate no real improvement is given by adding the wavelength into the equation; therefore, it may be profitably deleted.

If several wavelengths (variables) have low t or F values (less than 10 or 100, respectively), it may be advisable to delete each of the suspect wavelengths, singly or in combination, to determine which wavelengths are the most critical for predicting constituent values. When the data are highly intercorrelated, then wavelengths containing redundant or correlated information can be identified in this way. If such a wavelength is deleted, then the t values for the other wavelengths will increase markedly. In such a case, where an important wavelength is masked by intercorrelation with another wavelength, a sharp increase in the partial F (or t) will occur when an unimportant wavelength with high intercorrelation is deleted, since that reduces the high intercorrelation between the variables still included in the regression equation.

The t statistic is sometimes referred to as the ratio of the actual regression coefficient for a particular wavelength to the standard deviation of that coefficient. The partial F value described is equal to this t value squared; note that the t value calculated this way retains the sign of the coefficient, whereas all F values are positive.

7. *Statistic:* Standard error of estimate (SEE). Also termed standard error of calibration (SEC)
Abbreviation(s): SEE, SEC
Summation notation:

$$\mathrm{SEC} = \left(\frac{\sum (Y_i - \hat{Y}_i)^2}{N - K - 1} \right)^{1/2}$$

Computational formula:

$$\mathrm{SEC} = \mathrm{MS}_{\mathrm{res}}^{1/2}$$

Comments: This statistic is the standard deviation for the residuals due to differences between actual (primary reference laboratory analytical) values and the spectroscopically predicted values for samples within the calibration set. It is an indication of the total residual error due to the particular regression equation to which it applies. The SEC will decrease with the number of wavelengths (independent variable terms) used within an equation, as long as the added terms have real modeling capacity; this indicates that increasing the number of terms will allow more variation within the data to be explained, or "fitted". The SEC statistic is a useful estimate of the theoretical "best" accuracy obtainable for a specified set of wavelengths used to develop a calibration equation. The SEE is the statistician's term for SEC (which is the spectroscopist's term for the same quantity), both designating the square root of the mean square for residuals. In this calculation, the residual for each sample is equal to the actual chemical value minus the spectroscopically predicted value for all samples within the calibration set.

8. *Statistic:* Standard error of prediction
Abbreviation(s): SEP (used to indicate standard error of prediction, or standard error of performance), SEV (standard error of validation)
Summation notation:

$$\mathrm{SEP} = \left(\frac{\sum (Y_i - \hat{Y}_i)^2}{N} \right)^{1/2}$$

Computational formula: none
Comments: The SEP stands for the standard error of performance, also known as the standard deviation for the residuals due to differences between actual (primary wet chemical analytical values) and the spectroscopically predicted

values for samples outside of the calibration set using a specific calibration equation.

The SEP is calculated as the root mean square difference (RMSD), also known as a mean square for residuals for N degrees of freedom. It allows for comparison between NIR observed predicted values and wet laboratory values. The SEP is generally greater than the SEC, but could be smaller in some cases due to chance alone. The crucial difference between this statistic and the SEC is that the samples used to calculate the SEP must *not* have been included in the calibration calculations. When calculating the SEP, it is critical that the constituent distribution be uniform and the wet chemistry be very accurate for the validation sample set. If these criteria are not met for validation sample sets, the calculated SEP may not have validity as an accurate indicator of overall calibration performance. To summarize, the SEP is the square root of the mean square for residuals for N degrees of freedom; where the residual equals actual minus predicted for samples outside the calibration set.

9. *Statistic:* Standard error of cross validation (SECV)
Abbreviation(s): SECV
Summation notation: (see SEP—Statistic 8)
Computational formula: none
Comments: The calculation of SECV is the same as for SEP. The difference is that this statistic is used as a method for determining the "best" number of independent variables to use in building a calibration equation, based on an iterative (repetitive) algorithm that selects samples from a sample set population to develop the calibration equation and then predicts on the remaining unselected samples. Some procedures for calculating SECV may calibrate using from almost all to two-thirds of the samples while predicting on the remaining samples. The SECV is an estimate of the SEP and is calculated as SEP or SECV as the square root of the mean square of the residual for $N - 1$ degrees of freedom, where the residual equals the actual minus the predicted value. In many statistical software packages, the SECV is computed for several different equations; the equation with the lowest SECV being selected as the "best" calibration.

This method for testing calibration models is often used for multiple linear regression (MLR), principal components analysis (PCA), and partial least squares (PLS).

In the limit, this algorithm grades into the PRESS algorithm, where only a single sample is removed from the sample set for calibration, then the removed sample is predicted. This cycle is repeated for every sample in the calibration set (the removed sample is returned to the calibration set when the next sample is removed); then the PRESS (prediction error sum of squares) value is computed as the root mean square of the errors due to each removed sample.

10. *Statistic:* The bias-corrected standard error
Abbreviation(s): SEP(c), SEV(c)
Summation notation:

$$\mathrm{SEP} = \left(\frac{\sum (Y_i - \hat{Y}_i - \mathrm{bias})^2}{N - 1} \right)^{1/2}$$

Computational formula: none
Comments: Bias-corrected standard error measurements allow the characterization of the variance attributable to random unexplained error, with the systematic (constant) difference removed. This statistic is useful since, if there is in fact a bias between the predicted and actual values, the calculated value of the SEP will be inflated by the bias term. Since the bias is easily removed by simply adjusting the b_0 term of the calibration equation, it is useful to be able to assess the bias-corrected accuracy of the calibration with equal ease. The bias value is calculated as the mean difference between two sets of data, most commonly actual minus NIR predicted values.

11. *Statistic:* Standard deviation of difference (SDD)
Abbreviation(s): SDD, $\mathrm{SED}_{\mathrm{duplicates}}$
Summation notation:

$$\mathrm{SDD} = \left(\frac{\sum_{i=1}^{N} (Y_1 - Y_2)^2}{2N} \right)^{1/2}$$

Computational formula: none
Note: If either $Y1$ or $Y2$ is considered to be the "true" analytical value, then using N in the denominator is the appropriate measure of the error due to the other method. However, take good heed of our warnings regarding untested assumptions!
Comments: Standard deviation of difference (SDD) or standard error of differences for replicate measurements ($\mathrm{SED}_{\mathrm{replicates}}$) is sometimes referred to as the repack error, calculated to allow estimation of the variation in a spectroscopic analytical method due to both sampling and presentation errors. It is a measure of precision for an analytical method. The precision measure can include multiple physical sources of variation of the data; therefore, this calculation is profitably used in conjunction with nested experimental designs with multiple levels of nesting.

Since real measurements include error, and this flies in the face of the fundamental assumption of regression analysis (that there be no error in the X variables) this statistic is useful as a means of comparing the effective error in X with the total error (or error in Y) to assess the relative effect of the X error.

The same calculation can be applied to replicate measurements of the reference method, to assess the accuracy of the reference laboratory.

12. *Statistic:* Standard error of the laboratory (SEL) for wet chemical methods (i.e., the method against which the instrument is calibrated)
Abbreviation(s): SEL
Summation notation:

$$\mathrm{SEL} = \left(\frac{\sum_{j=1}^{M} \sum_{i=1}^{N} (Y_{ij} - \bar{Y}_j)^2}{M(N-1)} \right)^{1/2}$$

Computational formula: for pairs of readings, the formula may be simplified:

$$\mathrm{SEL} = \left(\frac{\sum_{j=1}^{M} (Y_1 - Y_2)^2}{2M} \right)^{1/2}$$

Comments: In the above formulas, M represents the number of different samples included in the calculation, while N represents the number of readings taken on each sample. This statistic is effectively a computation of the random error (precision) of the reference method; the values of Y represent various determinations of the constituent composition by the reference method. There are a number of ways to compute SEL. To compare SEP to SEL, the SEL can be determined by using a single sample properly split and analyzed in replicate by one or more laboratories; then the standard deviation of those readings may be computed and used. The difficulty with this is that it does not take into account the possibility that different samples may have different variabilities (say, due to different degrees of inhomogeneity). Alternatively, many different samples may be each read as few times as twice each. In either case the formulas are valid with the proper values used for M and N. In a good calibration, the SEP is generally 1.0–1.5 times the SEL.

These formulas can apply whether the replicates were performed in a single laboratory or if a collaborative study was undertaken at multiple laboratories. Additional techniques for planning collaborative tests can be found in Reference (37-1).

References

37-1. Youden, W.J., and Steiner, E.H., *Statistical Manual of the AOAC*, 1st ed., Association of Official Analytical Chemists, Washington, DC, 1975.

38

THE BEGINNING ...

(Author's note—most of the columns that originally appeared in *Spectroscopy* had to be converted to be in a suitable format for book chapters. This concluding chapter, however, could not have been converted without losing a good deal of the essence of what was being said; consequently, some sections were left unchanged, so if they read funny, as if they don't belong in a book (such as the very next paragraph) our readers will know why).

"We're guessing, of course, because at the time we write this there is a considerable backlog of columns to be printed before this one comes out, but our best guess is that by the time this one does actually appear in *Spectroscopy* it will have been about five years since we first embarked on this project." was the original introduction to this paragraph. In this second edition, we now know that the last column to be included in this book was printed in *Spectroscopy* in Volume 8, Issue 1: the January, 1993 issue of *Spectroscopy*. In that time (i.e., since the beginning of this book—author's note) we have covered a good number of topics, interspersing spectroscopy and whatever mathematics or statistics we thought we could get away with, with explanatory and even some philosophical discussions of what statistics is and can be used for.

Starting with some basic principles, we progressed to discussions of some of the elementary statistics that are commonly encountered in dealing with data.

More importantly, we tried to bring out not only what the statistics *are* (which almost anybody can—and usually does—do) but also how they are used to describe the behavior of data (which is normally done only by and for people who call themselves statisticians—there is nothing wrong with being a statistician, by the way; if you have been paying attention, then you know why statisticians exist and why there is a crying need for good ones). On the other hand, those of us who think of ourselves as something else (chemists, for example?) still have need for some of the knowledge that statisticians are privy to. There is far too much of this knowledge for us to have included more than a small part of it in our book. However, if we have done the job we set out to do (which was to present what information we could in a way that is understandable enough for our readers to be able to follow along and interesting enough for you to *want* to follow along), and you have done the job we set out for you to do (which was to read what we wrote with enough attention to actually absorb some of it), then by this point you should be able to pick up just about any book on elementary statistics (such as the ones we presented and recommended from time to time) and to keep up with the more advanced discussions of the topics covered.

In at least one respect, statistics is like chemistry. The field of study we call "chemistry" is divided into many subfields (analytical, organic, inorganic, physical, biochemistry, etc.). Just as a chemist starts out by taking a course in "general chemistry", where a brief overview is given to some of the topics, and then takes specific courses in various of the more specialized subdivisions, so too does a statistician learn his field by first learning a little bit about a broad range of topics and then studying various selected ones (e.g., ANOVA, regression, etc.) in greater depth. We have to some extent tried to follow this format, although at a more elementary level than someone who thinks of himself as a statistician would be satisfied with. Thus, in a manner similar to that by which a chemist studying analytical chemistry, for example, has acquired background knowledge of pH and phase changes (to pick two topics at random), so too does a statistician have a background knowledge of probability theory and descriptive statistics when studying, for example, analysis of variance. It is this smattering of background that we have tried to impart, for two reasons: first, so that you can recognize when one of these subdisciplines within the field of statistics can be helpful to you; and second, so that you will be able to recognize what is going on when you do try to delve a little deeper into any of the subfields, and not be chased away either by unfamiliar terminology or scary-looking equations.

Despite statistics being considered a mathematical science, at a certain level very little mathematics is actually needed. To understand the basic concepts of the effect of variability of data on the measured values from the calculations based on those values requires no math at all. A good deal of the math that is actually useful (aside from simple arithmetical calculations of mean and standard deviation and so forth) is all lumped together in a partitioning of sums of squares. Once you understand how that is done, you have actually licked about 70 or 80% of the problem. Besides, grasping the concept of how sums of squares are partitioned is such a gee-whiz thing that it boggles the mind (well, our minds, anyway—but then some minds are more easily boggled than others) and makes it completely impossible to understand how anyone can *not* pursue this topic to the point at least of comprehension.

Having reached that point, it is relatively straightforward to go on and study more advanced topics, and more advanced treatments of them (note that we said straightforward, not necessarily easy), whichever ones turn out to be important for you. You will have laid the foundation for being able to tackle these topics with assurance of being able to follow what is going on.

Which brings us to the title of this chapter. "The past is prologue..." Prologue to what? Well, here we come to the "C" word. Anyone who is afraid of the "C" word should close their eyes before reading the next section, and not open them again until they have finished it. The current buzzword in treatment of data from analytical instruments, especially spectrometers of various sorts but also chromatographs and to a lesser extent other instruments, is chemometrics (there, we have said it, and in public, no less). We do not know offhand if there is a formal, official definition of chemometrics, but for our purposes it can be considered to be the application of multivariate techniques to the analysis of data of chemical interest. In fact, we have already presented at least one such: multiple regression, used as a tool for determining the calibration coefficients for spectrophotometric analysis, fits the definition perfectly. Strangely, it is sometimes not included in discussions of chemometrics. Our best guess for this is because it is an "old" technique, whereas the term "chemometrics" is reserved for newer, more "modern" methods.

On the other hand, because regression is an older technique, it has been much more thoroughly studied than the "modern", "chemometric" methods, therefore its behavior, and particularly its weaknesses, are much better known. In particular, methods for determining the sensitivity of the results from regression analysis to the variability of data have been worked out, and so, for example, the vast majority of multiple regression programs include the calculation of the auxiliary statistics that indicate how variable the results are.

Many of the modern techniques are more widely discussed, and sometimes touted as being the "last word" in solving all manner of spectroscopic analytical problems—the "magic box" into which you put some data and out of which comes an answer. A good deal of work has gone into the extensive (and difficult) mathematical development of the many chemometric techniques that have become available, and we do not want or intend to minimize the progress that has been made. However, because of the newness of these mathemagical methods, little attention has yet been paid to such items as the variability of the results, i.e., the multivariate equivalent of the variance of the variance (as we called it in Chapter 17).

Understanding univariate statistics, such as we have been (mostly) presenting, is one key step in proceeding further.

A useful analogy can be drawn between some of the univariate techniques and multivariate methods. We have seen how some of the techniques are inherently continuous (e.g., calculations of means and standard deviations of continuous data), while some are inherently discontinuous (the statistics based on counting). This shows that discontinuous events, which are also subject to random influences, also give rise to data that can be described by some sort of mathematical discussion. There are multivariate analogs to this: two methods of qualitative analysis using spectroscopic data are, for example, discriminant analysis and nearest-neighbors analysis; these represent the continuous and discontinuous cases, respectively.

The problem is that even in those cases where an analytic mathematical description has been generated (which is no mean feat in itself, and beyond the capabilities of your friendly local commentators), in general the results are so complicated and difficult to use that they are essentially never implemented in conjunction with the algorithms that are provided to do the main data analysis task. This vacuum leaves the implication that no further variability is involved; that the main algorithm describes the data exactly and that is the end of the story. Of course, by now we should all know better.

So how can we deal with this situation? Multivariate data have characteristics that univariate data lack (e.g., correlation: one variable does not have a correlation since this requires at least two); this prevents us from creating oversimplified examples such as we were able to do for the simpler univariate statistics. However, we do plan to do two things: first we will discuss some chemometric methods in a relatively straightforward way (one might say that we will use the "classical" discussion), but by giving it our own special twist, and by keeping it as simple as possible, and by using our own examples and insights to try to clarify what is going on.

Secondly, we will try to introduce the concepts of variability of the quantities involved. Partly this will be done through the use of computer simulations (such as one previously reported (38-1) and about which we will have more to say), and other research techniques; indeed, we may actually be doing some exploratory research as we get deeper into the subject. This may get a bit tricky—it is tantamount to publishing research results without peer review—so we will rely on our readers to keep us honest. On the other hand, if you really follow along and keep up with our provocative (we hope) "exercises for the reader", then you will also be part of our research team (audience participation research?) and can share the credit (or blame, as appropriate—after all, you will all be accessories after the fact).

So this really is a beginning. Until now we have been trampling all over the topics that are pretty thoroughly explored, and that only needed to be described. We have discussed what we could, and in our own minds, at least, finished what we set out to do. For the benefit of those of our readers who may have missed some columns and wish to fill in, as well as for some other reasons, we have collected all the columns printed until now and are publishing them as a book [available from Academic Press by the time you read this (exercise for the reader: guess the name of this book!)]. We now embark on a new phase of this column, where the subject matter is not nearly so well known, much more complicated, and in some cases, may be unexplored. We may do some floundering in places that comes with the territory. Judging by the reports we receive from Mike MacRae (the editor of *Spectroscopy*, for the benefit of those of you do not read the credits—Mike, take a bow) the reader service cards indicate that this column is pretty popular. We hope we can do as well when we extend the column into the new territory, and start discussing... (Exercise for the reader: what do you think the new name of the column will be?)

(Author's note—Well, that's what we wrote for the transition of the column from discussion of statistics, to the discussion of chemometrics. This book must end, and this transition point seems an appropriate place, but the study of data doesn't. What we learned from our statistician friends is that what you can see in data depends upon how you look at it. We've previously mentioned "the art of science"; in this case it means learning how to look at data in order to extract information. There are many ways to do this, of which chemometrics is only one; nevertheless, it is one that can benefit from some of the tricks that the "old masters" discovered and used.When our columns about chemometrics become sufficient, we will collect them also, and release another book. In the meanwhile, you can still get them free, as columns in *Spectroscopy*, although if the scheduling of that book comes out the way this book seems to be going, you'll still have to buy that one to get a preview of how it comes out).

References

38-1. Mark, H., *Applied Spectroscopy*, 42(8), 1988, 1427–1440.

INDEX

A

V

W

X

Y

Z